# Quin Kola:

## Tom Payne's Search for Gold

# Quin Kola:

## Tom Payne's Search for Gold

In which friends, acquaintances, and family members recall his adventurous life

Alice V. Payne

To Al,
with best wishes,
Alice V. Payne

Canadian Cataloguing in Publication data

National Library of Canada

ISBN 0-9686646-0-1

1. Canadian Mining History
2. Geology
3. Prospecting
4. Alberta Petroleum History
5. Biography

Alice V. Payne, 1940-

Cover Painting by Maurice Haycock (1900-1988), Geologist-artist:
**Looking Across Echo Bay Towards Mystery Island (1950)**

Book and Cover Design by Harald Kunze

Illustrations by Larry Stilwell

Edited by E. Tina Crossfield

Printed by McAra Printing, Calgary, Alberta, Canada

Published by Crossfield Publishing, Okotoks, Alberta, August 2000.

To Olga
who made Tom
write it down and
to all those prospectors
who follow a dream

# Contents



# Maps

# List of Illustrations

# Notes to the Reader

Imagine that you have just pulled up a chair by the fireplace and are cradling a mellow glass of Dewar's Scotch Whisky. As usual, Tom Payne's tales of the north have captivated you, and he is holding center stage among the gathering. A number of his acquaintances are in attendance, and they interrupt him, anxious to add their recollections to the general conversation. The talk is an entertaining mix of frontier towns, prospecting days, and oil booms. Some stories are strangely familiar. Others are new and exciting, and leave you with vivid images and always more questions.

My father lived a life following his dreams. In 1964, after two operations, he recuperated by writing his memoirs. He called it his diary, and shortly before his death, he sent it to me with instructions to keep every scrap of it intact. The two notebooks are located in the Public Archives of Canada. I checked most of the names and dates and found them to be surprisingly accurate.

This book is based on Tom Payne's journals, supplemented with the oral history of a wide variety of friends and associates. As a geologist, I wanted to bring my father's prospecting activities to life for the public, complete with the hardships, adventures, and joys of discovery. Today there is still room for the lone prospector assisted by persistence and the art of chance. The recent diamond and nickel finds in the north confirm this.

I gleaned background information from mining journals, company records, newspapers, magazines, and history books, and used private family papers, letters, and diaries. I was fortunate in being able to interview a number of friends, relatives, and business associates who knew my father personally. Although there are gaps for which no data exists, the source materials have remained largely intact.

The unusual format allowed me to explore and expand beyond conventional biography. Source materials have been edited only for style and flow. Short biographies of the speakers appear at the end. There is no definitive voice here. Rather, this collage has been lovingly assembled and retold as he might have wished.

Information and valuable assistance was provided by the libraries and staff of the Geological Survey of Canada, the Northern Miner, the Public Library and Archives of Canada, the Glenbow-Alberta Institute, the Western Development Museum, the RCMP, the Hudson's Bay Company, and in England, Felsted School, and the Chelmsford Public Library.

Cominco supplied maps, meals, and transportation in Yellowknife, and Gulf Canada Resources furnished me with office support in Calgary. Bob Hauser of the Miramar Con Mine kindly provided a surface tour of present operations. The Alberta Chamber of Resources, the Canadian Institute of Mining, Metallurgy and Petroleum, and the Petroleum Society, the Canadian Society of Petroleum Geologists, and the NWT Chamber of Mines provided additional help.

I am very appreciative of financial support from the Canadian Geological Foundation to aid in publication.

Illustrations come from a variety of sources. Stan McCormick, Gordon Willsie, Lyman Skinner, and Robert Folinsbee donated many of the historical photographs. The rest are from my own collection. Cominco and the Geological Survey of Canada supplied several photographic references. Ian Folinsbee redrafted all of the maps, and Larry Stilwell created all of the line drawings.

I wish to thank Karole Pittman and Kathy Haycock for their permission to reproduce the Maurice Hall Haycock painting

*Looking Across Echo Bay Towards Mystery Island* that appears on the cover. Their efforts in transporting the artwork for digital photo imaging were most appreciated. Maurice Haycock was a geologist who had a keen interest in art. His style, influenced by members of the Group of Seven, reflects the landscape as it was when Tom Payne was prospecting in the north.

A heartfelt thanks goes to Jed Dagenais, Robert Folinsbee, Elizabeth Macpherson, Linnet Murray, Veronica Pounder, and Sue Rose who read the manuscript and offered valuable suggestions. Katherine Payne commented on the first version of the manuscript, and Cathryn Gunn acted as copy editor for the final text.

I would also like to thank Tina Crossfield, my patient and thoughtful editor, for her enthusiastic guidance. Her helpful suggestions and persistence have made this book a reality.

A final thank you goes to all my family, the cheering squad whom I have occasionally abandoned while I worked on this project. I hope it will provide everyone with information, inspiration, and entertainment for years to come.

# Prologue

I was damn near snow-blind and so was Stockings, my Inuit partner. After finishing a long freight-haul with caterpillar tractors, we'd gone trapping. Coming down to the coast for supplies, we got caught in a real bad storm.

When we reached Eskimo Point, north of Churchill, Stockings took our furs to the trader, an old Scotsman who lived in a rough dugout on a hillside. It was just a shack with a pole front and a table for bartering. The trading goods were stacked against the walls behind him.

We had three bales of lovely white fox, and that trader wouldn't give us a thing! He told Stockings to bring them all inside, and then they would settle up outside. I knew exactly what that meant: he'd have the furs and shut the door, and we'd be left cold and starving, standing in a snow bank.

So I went back alone and laid my .30-30 on the bench. The trader asked, "What have you got that for?" I replied, "I'm going to use it, and use it plenty. This gun is loaded and I'm half-dead already, so it makes no difference to me if I kill you or not! So get busy. This is what I consider my furs to be worth." I pointed to some flour, baking powder, tea, and sugar. I carried it all out to Stockings who was so surprised, I thought he'd die anyway!

In the meantime, Stockings had made an igloo. It only took him forty minutes. We cooked up a stack of hot cakes about a foot high and tended to the dogs, even feeding them the leftover fried bits from the pan. Then we cached our groceries in the entrance-way along with our furs, and closed the hole with a block of snow. The dogs took their hunks of crackling and burrowed into the drifts outside. In ten minutes, they were invisible.

Being snow blind feels just as if someone threw a handful of sand into your eyes. You pretty well have to go to bed for a week or two but you can't if you have nothing to eat and it's forty degrees below zero. You can't light a Primus stove very well when you can't see. So you make tea, and lay the bags on your eyes. They run with ooze and glue shut. When the crust cover comes off, it lets the light back in. We couldn't get up, so we took the hot cakes to bed with us so they wouldn't freeze. No heat, no nothing, just bed. We had to be damn careful, and finally we got better.

Boy, that was something! That dirty trader was a real son of a bitch! But it wasn't his fault. He only got paid about $25 a month. The company would bring guys like him from the docks of Glasgow right to the shores of the Hudson Bay, drop them off, and pick them up the next year. If they lost a knife, an axe, a rifle, or anything it came right off their paycheck. Most of the poor suckers couldn't even read or write.

When you've had it done to you, you remember. It was tough walking across the barren lands with nothing but a rifle and a tarp, and living off the land. I know what it is to be broke and hungry, and to see the dogs called into the cook shack and fed leftovers for fear that I would eat off their slop pile. After I staked one of my best claims, I was kicked out of camp and had to sleep under a tree root because I wouldn't sell.

I know what it is to be dying in the hospital under an oxygen tent, and have vultures stand by my bed and say, "Tom Payne, you are going to die tonight. You owe the hospital and the doctors a thousand dollars each. What about giving me ten percent of your interest in Ryan Gold Mines and I will settle up that little debt for you?" My reply was if they couldn't pull me through then it was too bloody bad for them!

That was a desperate situation but I survived, and the mine prospered. So did all the shareholders. But not all my efforts succeeded. The best looking gold find I ever made was at Courageous Lake, but we drilled it and it bottomed out. That was a terrific disappointment, but I can't complain! One gold mine, one oil well, and one good family are enough for any man in one lifetime.

Thomas Payne

# English Roots

## Chapter 1

Alice Payne    Dad was born in Witham, Essex County, on January 11, 1891. He was the eldest son of Dr. Frank Payne and his wife, Alice Harvey. Her relatives were flour millers. As a young girl, she went to live with Marian and Dr. Tomkin who had no children of their own. The public version of the story is that Dr. Tomkin met Alice at the age of three, was captivated by her charm, and arranged for her adoption. Privately, Alice was believed to have been his illegitimate daughter.

Dr. Payne, my grandfather, received his medical degree from Cambridge and arrived in Witham in 1886. He was a member of the Royal College of Surgeons but had neither money nor a practice until he met Dr. Tomkin and became his assistant. Dr. Tomkin had inherited the busy practice from his father, along with a private lunatic asylum called The Retreat, located just down the street from his residence at High House.

Alice Harvey was engaged to a solicitor's son from Colchester, but she changed her mind after meeting Frank. They married in 1889 and settled into High House with Dr. Tomkin. After his death, grandmother inherited his entire estate worth more than £23,000. Frank took over the medical practice and the lunatic asylum, which he visited daily. The family soon grew to include my uncles Henry and Bill, and aunt

Marian. Along with Annie Smith, the nanny and housekeeper, my grandparents employed several servants, and a groom for the horses.

High House was an impressive property with a large garden, and eight acres of meadow. It was furnished with Chippendale furniture, Rembrandt paintings, and enough Queen Anne silver to fill three barrels. In the basement was one of the best wine cellars in Essex. The landholdings encompassed five cottages and a few more houses in Witham.

In 1898, about the time Dad got his first pair of long pants, the family was implicated in a series of lawsuits that intrigued and entertained the whole county. The Paynes saved the newspaper clippings of the trials, along with artist's sketches, in a large black scrapbook. All of us still talk about these events, which exposed the family to public scrutiny and ridicule.

Mr. Clements, a greengrocer, who alleged that grandfather had assaulted his thirteen-year-old daughter, brought the first case. The girl maintained that Dr. Payne pushed her down and swore at her as she walked to the sweet shop. Her brother, age eleven, confirmed the story. Mrs. Clements, who was upset about the mud on her daughter's dress, took her to see another doctor who admitted that he'd found very little wrong. At the end of the testimony, Mrs. Clements said that her brother Mayhew, a fishmonger, had urged her to lay the complaint.

Of course, grandfather denied the whole story. He hadn't seen the girl, and the first he'd heard of the matter was when Mayhew and Mrs. Clements had come to his house. Mayhew stated his business and grandfather told him to get out. After a violent argument, Dr. Payne telephoned the asylum for an attendant and threatened to lay charges. In the end Mayhew left, no doubt thankful that he hadn't ended up in a straitjacket.

The defense lawyer suggested the girl had made up the whole event when her mother scolded her for getting dirty, and observed that the matter had only come to light because of Mayhew's meddling. The case was dismissed amid applause.

Within three months the Paynes were back in court. "Wickham vs Payne" dominated at least twenty local and London newspapers for two days: "Remarkable will suit - A dead doctor's money - Allegations of undue influence - Extraordinary statements!"

The case was heard in London before a judge and jury. This time, Mary Wickham, a widow and sister of Dr. Tomkin, sought to have his will set aside. The trial promised to be sensational: a rich doctor's will, his disinherited sister, and an unscrupulous son-in-law. There was talk of blackmail, murder, and exhumation of bodies. The court was crowded with barristers, witnesses, reporters, family friends, and of course, Mayhew, the fishmonger.

The story went something like this: when Dr. Tomkin died, my grandparents had been married for six years. Now, four years later, Mrs. Wickham was insinuating that he'd been poisoned. Wickham reasoned that her brother had broken Alice's first engagement, then forced her to marry Dr. Payne against her will. Apparently seized with guilt, Dr. Tomkin left Alice most of his fortune, thereby excluding his dear sister, who only wanted justice (not the money). If that wasn't enough, she implied that Dr. Tomkin had performed an illegal operation upon a young woman, giving Dr. Payne a motive for blackmail. She insisted that a guard be placed upon her brother's grave, and tried to get his body exhumed.

In fact Dr. Tomkin had made several wills, all of which named Alice as the principal beneficiary. He'd already left funds to his sister, but her portion was revoked after she had involved him in a chancery suit charging him with being a fraudulent trustee. She'd also spent more than £20,000 and had declared bankruptcy. To the shock of her attorney, the defense lawyer produced a series of intimidating letters she'd written demanding money and threatening legal action. After a short recess, the defense asked the Paynes to deny these charges and the judge dismissed the case. By April, everyone was back in court.

The third and last trial was billed by *The Times* as the sequel to the famous inheritance suit. This time, "Payne and Wife" were the plaintiffs and they were suing Mayhew, the fishmonger, for slander. Town gossip suggested that the Paynes had murdered Dr. Tomkin, and that Frank was a habitual drunkard and had defrauded the Queen by making false tax reports. It was even whispered that Dr. Payne had been found in Mayhew's daughter's bedroom!

Mayhew kept insisting that he was innocent and therefore couldn't apologize for something he hadn't done. Although the Paynes didn't wish to ruin him, they did want public vindication of their characters. It was a stalemate, and the judge adjourned the case without a verdict.

The charges of blackmail, murder, and assault only obscured the minor notes about grandfather's conduct in the community. Because these major sins could not be proved, the lesser ones were ignored. On the witness stand, he was the wealthy, successful doctor: a man of letters and integrity who made a good impression on the witness stand. He could also be completely captivating and charming to his friends. Out of court, he was easily offended and had a violent temper. Some of the charges made during the trials came true in his later life: he could be brutal in his treatment of others, family or not.

DAVID PAYNE I was told that grandfather had been an alcoholic, but some scare stopped him and all he ever drank later was a glass of port. His hobby was growing large dahlias in a conservatory when he lived in Clacton. When I was little, my mother used to take my sister and me there to visit. On leaving, we would both be called to his chair and given a bit of cash. Always an odd amount, like three shillings and seven pence. God knows what we did with it, I never remember buying anything.

TOM PAYNE He rarely gave me any spare change, and we certainly didn't get along. Everything I did irritated him and vice versa. For a doctor, he could be terribly cruel. We had a root house in the back yard, full of potatoes for the winter. Someone had been stealing the odd sack, and my father decided to catch him. In those days, you could just put up a sign like "No Trespassing - Beware - Spring Gun and Man Trap Ahead!" Perfectly legal. If you killed or maimed someone, it was considered his own damn fault.

Along with his sign, my father set out a bear trap, one of those large black iron ones. Further down the trail, overgrown with grass and weeds, he also set up a spring gun. Anyone coming along would trip the wire. Well sure enough, the poor bugger who had been stealing potatoes ran right into the bear trap. It nipped off his leg at the knee. He lost a lot of blood and nearly died from shock. My father bandaged him up and sent him home, telling him he was lucky he didn't get shot!

He didn't treat his patients much better. One woman, who was a Christian Scientist, wouldn't take pills or tonics of any sort. But when she got bronchitis or pneumonia and thought she was about to die, she'd get scared and call him. One night, her husband telephoned the eminent Dr. Payne, but he refused to go. When the coroner asked why at the inquest, he explained, "If she died I'd get the blame, but if she lived, God would get the credit."

'TOMMY' PAYNE Dad could be mischievous too. When he was still a boy in Witham, he watched an organ grinder with a trained monkey who frequented the High Street. During the day, it would steal anything shiny and give to his owner while he kept playing. Dad lit a fire in the stove and heated a large penny until it was red hot. Then he placed it on a pottery saucer on the windowsill. Seeing the nice shiny piece, the monkey grabbed it, burned its hand, and went howling back

*Tom Payne as a young child.*
*Payne family collection.*

to the organ grinder, who cried foul and demanded compensation. Grandfather didn't pay for the monkey, but Dad sure got tanned.

ALICE PAYNE Fortunately, he was usually safely incarcerated in boarding school. The first school was Maldon Preparatory. It was far enough away from Witham to keep him out of his father's way, but near enough to walk home regularly to see his mother. He was nearly expelled as a result.

TOM PAYNE The School Headmaster wouldn't let me go the fourteen miles to see mother who was dying of cancer. So one night I escaped and detoured to Braintree and the White Hart Hotel. I got in touch with Jolly Old Taylor, the owner and a friend of the family. He fed me roast beef and gave me a bottle of Bass's ale before I set out for Witham.

Taylor was a fine man whose life ended tragically. He gambled on a horse race called the Cesarewitch, a handicap race run at Newmarket, near Cambridge. It was an important event, named after the Russian

*Tom Payne at age seven, posing in his first pair of pants before his departure to Maldon Prep School. Payne family collection.*

Crown Prince who had made a state visit to England in 1839. Taylor's horse failed to win and he lost £100,000! It cost the hotel everything. But he paid every cent back to his customers, then went to Clacton-on-Sea and jumped off the pier. Dr. Harrison, the coroner and Taylor's lifelong friend, brought in a verdict of 'found drowned,' although he secretly knew better.

I hid across from High House, after dodging the headmaster's Renault along the way. My father habitually went for his morning shave at the barber, and then to the pub for his whisky and cigar before lunch. I waited my chance, ran into the house, up the stairs, and went into my mother's bedroom. She was very ill, but her mind was clear.

I was very shocked to see her in that condition, but she did her best to comfort me, God bless her soul. She told me that father had promised to send me to public school for three years, and if I was ever in need, I was to find my Uncle Arthur at Birdbrook, in Essex. He proved to be pretty tightfisted. He only gave my aunt a pound per

*Tom Payne's sister Marian (age 12) and his younger brothers Henry (age 14) and Bill (age 5). Payne family collection.*

week to run the household and pay the hired girl. The rest had to come from game shot and trapped on the farm, and egg sales. Poor Aunt Ethel. The whole family pitied her.

Not all of my relatives were so frugal. Aunt Fannie, of Birdbrook, was a wonderful person. Every time we visited, she gave us kids a half sovereign. She'd become rich on rubber and tea plantations, and often financed her brother, Harry, our favourite uncle. Twice she saved the Eastern Counties Brewery for him, and three times he made a fortune. He would get ahead a few thousands pounds, acquire a fast woman and a few thoroughbreds, and they would go off to the Ostend racetrack.

Afterwards, there was hell to pay. He would wire Fannie and she would send his return fare. By this time the brewery would be in liquidation. So she would buy it back, and he would repay her. Then away he would go again. Harry never learned the lesson, and apparently neither did Fanny. When she died, he straightened up, well aware that if he went on another ramble he'd have no wealthy sister to bail him out.

*Felsted School, Essex. AERO Films, Ltd.*

Fannie never married, and she suffered from asthma in her old age. She used to take a small dram of belladonna and aconite before bedtime. She had a habit of making a new will almost every month with a new benefactor. When she finally pegged out, the coroner, Dr. Harrison, brought in a verdict of 'heart failure.' No friend of his was ever murdered or committed suicide! But I had a shrewd suspicion that Fannie's maid gave her an overdose. She stood to inherit a great deal, and knew that Fannie could change her mind again at any time.

ALICE PAYNE When grandmother Alice died in July 1905, Annie, the housekeeper at High House, cared for everyone. Dad, in particular, returned her affections and years later, sent her regular letters from Canada. She remained with Frank until his death in 1947. She inherited his bed, the covers and linens, four months of unpaid wages, and a bequest of £1000. A surprising and insulting end to many years of trust and affection.

*Alice Harvey Payne in 1905, sitting in the garden of High House shortly before her death. Payne family collection.*

Grandfather kept his word and sent Dad to Felsted School in January 1906. His introductory letter to the headmaster said that Dad was more fitted for a reformatory school than a public one! But the fees were low, meals were meagre, and dormitories were sparsely furnished. He disliked it intensely.

Felsted School still stands in a lovely village with a Norman Church tower, gardens, trees, and open spaces. Lord Riche founded it as a place of learning in 1564, and charitably provided for eighty boys. The aims of the school were carried out vigorously. The headmaster, Reverend Frank Stephenson, had an abrupt manner, expressed his opinions forcibly, and dispensed punishment severely and sternly. But he was a good teacher and got the best out of his students. Most of the boys began by disliking him, but ended with respect and admiration. Years later, Dad could still quote Latin and Greek with appreciation.

*Tom Payne, the Felsted Man, 1906. Payne family collection.*

Tom Payne A boy was either made into a real man or the system broke his spirit. Boys were caned for things like skipping compulsory exercises without a doctor's excuse, and other minor infractions. The black bruises on our skin were clearly visible in the swimming pool, and it was great sport in the dormitory pricking the blisters and letting the water out. We never wore short pants back then.

One day when the master was out of the classroom, a classmate poked me in the back with a compass, so I stabbed him with my fountain pen. He started to scream, which meant I was going to catch it. So I bet all the boys that I'd slip the lid of a Peak Frean's cookie tin inside my pants. I knew that the minute the cane sounded, I'd be in for extra swats, but it was worth £10 to me.

When the master came back, he sent me down to the library where they had a rack of canes: nice thick ones that don't hurt very much, and little whippy ones that snap and leave welts. On the first swat, there was a thwank that you could hear out in the hall. I removed the lid and collected my reward. Then we skipped out of school and went down to the local pub to celebrate, for which we got caned again!

I sat at the bottom of the class for three years, along with John Molson and Sammy de Beers. The latter always had a pocketful of uncut diamonds, and he used to pour them from one hand to the other. We'd play them like marbles when the master wasn't looking.

Alice Payne Near the school there was an abandoned railway line with an old steam engine. It was a tempting sight. Dad and his friends poured water in the tank and stoked up the boiler. When they finally got it running, no one knew how to stop it. The engine ran off the end of the tracks into the weeds and crashed!

Everyone knew that some Felsted boys were responsible, but Dad must have been a devil for punishment as he was the only one to confess. In those days, a boy in disgrace was literally drummed out of the school: the drummer's corps was lined up along the driveway, and the ex-student walked out the gate forever. Dad somehow avoided this humiliation, and left in the ordinary way in July 1909.

He had no clear idea what to do next. Under the terms of his mother's will, the family was to receive a regular stipend from the estate, while the rest was to be safeguarded for grandchildren. The trustees had the power to liquidate any of the assets providing the proceeds were reinvested and held in trust. In 1910, Dad and Henry both received £500. Henry, who was later killed at Gallipoli during World War One, left England to work on a tea plantation in Ceylon. Dad took a job in the London office of Strutt and Parker Limited.

This firm was involved in farming and real estate. They kept purebred Friesian cows, and sold milk at the wholesale and retail levels in London. As agents for the Earl of Leitrim's holdings, the Guy's Hospital Estates, and the blue-chip Anglo-Dutch plantations in Java, it was a powerful company. Working as a land agent, Dad dressed up in a black suit every day and took the train to London. One of his side duties was to enter horses in the races for Lord Dewar. The payment was a case of whisky, which he shared with his friends.

Dad lived at the Manor House in Terling with Mont Everard, who was hired at age seventeen by Strutt and Parker. As senior assistant, and then director, Mont was indispensable to their operations, which managed about sixteen farms and 14,000 acres. He was a charming and loyal friend, although he could be stubborn at times. They liked each other on sight. From Mont, Dad learned the techniques of mixed farming that would be so useful later on.

TOM PAYNE Mont was a shrewd bargainer and well known to auctioneers and cattlemen. Sometimes we would sail across the channel to sell our cattle. We would herd them down to Harwich and put them on the boat to Holland or Belgium. On the way over we would feed them straw and a little beer, and then sell them by weight. They would be so bloated up they could hardly move!

Mont was such a good judge of horses that he was appointed an official at the Irish Horse Show. His talent proved to be very useful. Instead of going straight back to England, we took in the thoroughbred races on the beach at Ostend and made enough money to gamble in the casinos. I was always on the lookout for my Uncle Harry! In the early morning, you could see the local police inserting a pound note in the pockets of the poor suckers who had lost everything, and committed suicide. No man was ever found without a little money in his pockets. That would have been bad for business.

DICK SMITH When Tom was broke, which was quite often, he lived with his friends. Everybody in the village liked him, especially mother who would always make room for him at the table. When the fair came to the village, he would give the school children free rides on the roundabouts.

Tom was one of the most charming, amusing, and well-mannered people I'd ever met. He made some lifelong friends at Terling, and stayed in touch with many of them. A great man for parties, he had a gift for playing the convivial host. Once, when he and his friends ran short of liquor, he decided to ride his bicycle to Witham, some four miles, to get a bottle from his father.

It was a wet and gloomy night, and when he didn't return, two of his friends went looking for him. About half way to Witham they saw his bicycle lying in the road. They found him in a pond up to his neck in water. When they hauled him out, he simply said, "Thank God I've got my Burberry on."

Tom enjoyed hunting, fishing, and riding horses, but he and his friends would occasionally let 'all hell' break loose. General elections were the best times. At night, they painted houses blue for Conservative, or yellow for Liberal, depending on the occupant's views. During meetings, they released sparrows with ribbons tied to them, and threw the odd inkwell. It always turned out all right, as the estate would pay the damages.

Townspeople said, and I believe it to be true, that Tom was responsible for firing the historic Chelmsford cannon. An Essex regiment had captured the old gun during the Crimean War. It stood outside the Shire Hall on a three-tiered granite plinth, pointing across the square and down Market Road. Children used to sit on it, but it was the open muzzle that fascinated Tom and his allies. Over several weeks, in the dark or while listening to the Salvation Army band, they loaded it with a carton of gunpowder, compressed it with paper, and laid the fuse.

On the designated night, a treacle tin filled with concrete was pushed into the barrel, and the trigger was lit in the wee hours of Sunday morning. The gun exploded with a terrific bang, hurling projectiles across the square. The debris hit the Golden Lion Hotel, and broke all the neighbours' windows. People rushed out into the streets to find out what had happened. No one was ever charged, and the damages were repaired, but the town permanently sealed the muzzle.

John Gardner    Tom was very handy with a shotgun and relished a good expedition. I used to invite him down to 'swell the bag' when Londoners of royal or noble blood came down to hunt Essex ducks and pheasants. We spent a lot of time wildfowling on the marshes of England's east coast, and polling our skiff with a four-bore shotgun mounted on the bow. He liked to stand quietly in wild places, sniff the salty weeds, peer for the skeins of mallards or geese, and listen for the rustle of wings overhead.

Alice Payne    Dad's home and family situation did not improve, and like many other young men, he became restless. He visited his brother Henry in India, and travelled to Russia. The idea of being a bank clerk, a bond salesman, or even an estate manager had very little appeal for a man of his energies and skills.

On practically every billboard throughout Europe, travel posters compared Canada to paradise. Toronto and Montreal were portrayed as being almost equal to Paris and London. A fine railroad linked the country and passed through some of the world's best agricultural land. Single men were encouraged to come as farm workers. Married ones could apply for their own land. Women were almost guaranteed a husband, or a suitable job as a domestic until better opportunities arose. Clear blue sky, and the scent of new mown hay, were free for the asking. It was a seductive picture.

Most men believed that life in Canada would be an extension of their comfortable lives, with a little adventure thrown in. No mention was made of the bleakness, harsh winters, terrible loneliness, monotony, cultural isolation, lack of medical care, and grindingly hard work that met most immigrants. They were unprepared for the frequent crop failures and the sight of solitary shacks sitting beside mosquito-infested sloughs. Dad was ready for whatever challenges lay ahead.

# Canadian Prairie

## Chapter 2

Jack Moar  Tom's experience was over so many different eras. I usually make it a habit with guys I meet in the mining business to ask them how they got into it, and he fascinated me. He lived through the opening of western Canada and the north. It's amazing that in one man's life, he accomplished so much. His abilities and persistence allowed him to cope with hardships that would have broken many others.

Bert Wheatley  You only got a little bit of Tom. He never sat down and told his whole life story to anybody. I knew him in the early days, before he started prospecting. He was a tractor man; that's how he made a living, and we became friends when we batched together. I was the cook on Tom's shift, but he could go all day without food. Often at suppertime, he wouldn't eat or say very much. Then we would go into town for the evening. Later at home, with the air cool and clear, he would announce he was hungry.

I'd put out a loaf of bread, a dish of butter, some celery and tomatoes, lots of cheese, the saltshaker, and half a case of beer. I never bothered giving him a knife because he'd never use one. Years later, when I saw him slicing his bread and 'good mouse trap cheese' it sure made me laugh! He would pace up and down the bunkhouse. Each time he passed the table, he'd break off a chunk, dip some celery in the salt, and just plain talk.

Tom Payne When I was still living with Mont, I met a stockman named J.D. McGregor from Brandon, Manitoba. His family had really prospered there, and they owned a hardware store and ran Glencarnock Farms. They had assembled one of the greatest herds of Aberdeen Angus cattle in North America. Later, J.D. became Administrator of the Yukon, a member of the Dominion Food Control Board during the war, and the Lieutenant Governor of Manitoba. He was the main reason I came to Canada.

Alice Payne The McGregors produced meat for local markets and the railway. By 1889, the completion of the CPR and the low freight rates allowed them to export cattle to eastern Canada and Great Britain. Because of his interest in breeding, J.D. won the Grand Championship in 1912 at the Chicago International Show, and repeated this feat the next year.

*Tom Payne, immigrant to the Canadian West, October 20th 1912. Payne family collection.*

He also operated the Southern Alberta Land Company, which was formed and financed by the House of Grenfell, a London stockbroking firm. They had obtained all the leases and rights to large tracts of pasture southeast of Calgary, amounting to a quarter million acres at a dollar an acre, providing it was irrigated.

The land was in Palliser's triangle and designated as arid and unfit for cultivation. An engineering plan, endorsed by British shareholders, envisioned the diversion of water from the Bow River at Carseland into a canal running south and east for 198 miles to Medicine Hat. A system of reservoirs, siphons, and other devices was to maintain the water supply during periods of low flow.

When he asked Dad why he wasn't in Canada already, he replied that it would be nice but he didn't have the money. But J.D. had made a big impression on him, especially with his stories of the Yukon. When J.D. said, "Here's £500. Go to Medicine Hat, report to the chief engineer and go to work," Dad took his advice.

'TOMMY' PAYNE Dad was scheduled to sail on the Titanic in steerage but fortunately missed the boat. He arrived in Canada on the next available passage in the spring of 1912. He looked around in Montreal, then fired up his motorcycle, and wearing his grubby sheepskin coat, drove clear across the country. It wasn't much of a bike to ride either, just a two bit, 100 cc engine on two old tires.

TOM PAYNE I worked on a survey crew for the summer, and earned enough to pay back my loan. J.D. was a shrewd business manager. As a boy, he'd shovelled manure with Clifford Sifton in his father's livery barn in Brandon. After Sifton became Minister of the Interior in the Laurier cabinet, he appointed J.D. as gold Commissioner of the Yukon.

When he got to Dawson City, he nailed shut the bars and dance halls, and later, re-opened them. Only this time, the merchants needed an operator's license. J.D. sold the licenses. The fee was small, but the prerequisites were large. One stipulation was that his brother, Bob McGregor, would be the distributing franchisee on the liquor sales to saloons. Both of them ended up with millions.

People said that in J.D.'s shack, you couldn't walk on the floor without getting gold dust in your shoes. Some of the fine gold would spill out of the pokes, or sample bags, right over the spot where he weighed them out. When he finally left town, the shack sold for a fortune!

The irrigation business was an unlucky venture. The company planted ten or twenty thousand acres to wheat, oats, and flax, but the yields were low. In 1914, it was so dry that farmers couldn't harvest even one bushel. When World War One came along, the money supply vanished. They began selling off company beef to the construction workers on the canal.

At the time, I was working for the company at Suffield and stationed at Ronalane, where they hoped to build a siphon across the Bow River. When I saw how things were going, I wrote to Mont and told him to sell his holdings immediately. No reply. Next, I wired him and demanded he get out of it or we were through! His reply? "Have reluctantly sold my shares in Southern Alberta Land Company, considered one of the best investments here in London." By July 1914, it was practically worthless.

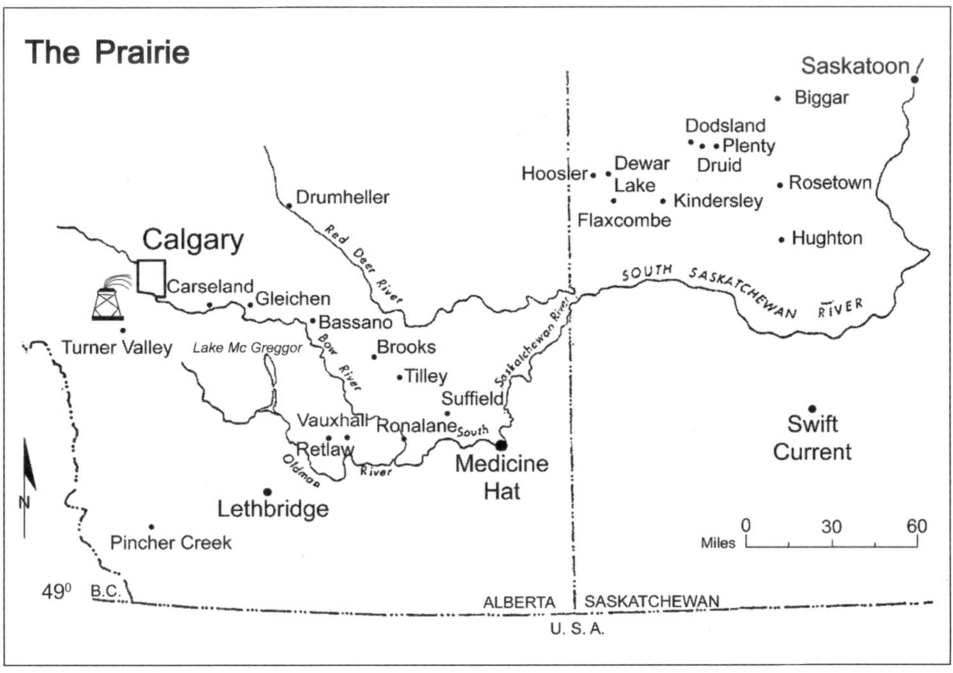

I left after it was liquidated to an American firm. A lot of Englishmen who were working there got fired. I hung on for six months as I was considered an American. But eventually my mail gave me away, and I quit. At the time I was running their warehouse for the winter, and being the last man in their employ, they told me to whistle for my money. This was all right by me! I had $1500 of theirs in the bank, and had the pleasure of settling with them and not them with me. Great jubilation on my part!

'TOMMY PAYNE' Dad had saved about twenty dollars to buy an old Model T Ford in Medicine Hat. He hoped to repair it, but it was just a pile of junk. Eventually he loaded it on a wagon and towed it with a team of horses. On the way to Calgary, he stopped at several farms for water and a chat. When a farmer asked if Dad had a spare radiator, he sold it. At the next few stops he sold the front axle, the spark coil, fenders, gas tank, seats, and all the bits and pieces that were removable. By the time he reached the city, he had one wheel left and $600 in his pocket, a lot of money in those days. With his profits, he bought himself another car, but couldn't afford to keep it!

ALICE PAYNE Young men like Dad, with respectable roots, and a boarding school education, often had little future back in England. They were called remittance men, or 'black sheep,' and many were given an allowance to leave home. These immigrants to western Canada caused much resentment among locals, and few people wanted to hire them. This damaging label persisted well into the 1920s.

HARRY HAYTER You just didn't want to have anything to do with Englishmen. Most came out here and thought we were a bunch of doughheads that didn't know anything. And they were going to tell us Canadians what to do?

TOM PAYNE After I arrived in Calgary, in 1914, I went down to the employment office. There were lots of jobs for compressor men in the new Canadian Pacific Railway (CPR) tunnel that ran through the Rockies. But the notice in the window said, "no Englishmen need apply." So I went in and confronted the operator. We got into a terrific fight. I knocked him down and dragged him into the gutter while a hell of a crowd collected. A chinook was on, and water was running right up to the curb. Then a policeman appeared and marched me off towards the jail.

He turned out to be George Gray, a remittance man from my hometown and brother of Leonie, my old girlfriend. Her mother, known in Brentwood as the 'Bank of England,' always gave Leonie a Rolls Royce, a blank chequebook, and all the champagne we could carry to the hunt balls and dances. Being a public school man, I had the entrée but not the means. My friend, John Gardner, and myself put on the first Bachelor's Ball at Colchester in order to make a few dollars, but it got too big and we had to drop it. Later it became a celebrated event held every year with tickets unobtainable.

I thought that George wouldn't recognize me, but after we were alone, he said, "See that lane ahead? When we get there, beat it!" I did, and I remember wondering what the hell to do next. The only two men I knew in Calgary were Bob Edwards and Pat Nowlan.

Bob Edwards was the colourful editor of the *Calgary Eye Opener*, and probably looking for an excuse to put the whole incident in writing. He was renowned for his defense of the downtrodden and helpless. Some of his best deeds involved a real estate man who had cheated a waitress named Rosie out of her life savings. Bob had gone to the man's office to ask for her money back, but was thrown out into the street. Next morning, the story, suitably embellished, appeared in the paper, complete with a short note promising that the scoundrel's name would be revealed in the next issue. Rosie got her money back in short order.

So I tore up to Bob's office and told him my tale of woe. "Here's $20," says he, "Go get yourself a room, get shaved up, buy yourself a bottle of whisky, and find yourself a good woman. The only stipulation is that you go down to the employment office and do it all over again tomorrow morning!"

I did, but the office wasn't open on Saturday. The next week, the front page of the newspaper read in big black letters, "Unknown Englishman beats up low down employment office operator." From that day onward, the sign disappeared from Calgary streets.

The warnings had been posted to curb remittance men who wanted extra income. Some got thousands of pounds a year from their rich families in England. Many had bought large ranches. A lot of them, God bless 'em, are still living around here and have made fortunes. Others depended on a small stipend that was sent to the local post office. Just enough so they could eat! The money was never forwarded, so they basically had to stay put.

When the employment office re-opened, I went back to see the CPR man. After three days of pestering, he decided to find me a job to get rid of me. He asked if I could handle a shovel, and gave me a note to take down to one of Fred Mannix's construction sites south of Gleichen. The man in charge said, "Oh, so you're the new shovel operator, are you?" I asked him for instructions, and he turned and pointed at a whacking great steam shovel out in the yard where they were back-sloping the hill to reduce the grade.

I had been thinking in terms of a hand shovel. But I went over and introduced myself. The operator tipped his hat and said, "I'm leaving tomorrow. Come on up here and I'll show you how it works. In a couple of hours you'll be as good as I am." I practised all night, swinging the bucket out, dragging it up the slope, and dumping the load. I ran that steam shovel for the rest of the season. Then the contractor got a gas-powered one, which suited me fine. I already had plenty of practice with motorcycle engines in England. We used to buy old beat-up machines, fix them, and then trade up.

When spring came, I hired out to Baldy Pearson, a contractor also known as 'Pick Handle Pearson.' What little hair he had was red, quite a contrast from his balding top. He was working on the Bassano Empress & Swift Current Railway, then under construction, and I hauled freight out to his camp near Tilley. I also acted as camp clerk.

One day at the halfway house, I lost two halters while driving a democrat, or light wagon, from Brooks. Pearson told me to deduct the $7.50 from my wages. That was a lot of money in those days. I was taking so much abuse from the foreman, and the high-line wagon skinner, I finally made up my mind to show them how hard it was to earn that cash.

Baldy never kept much plow point steel on hand to sharpen the grader, and often rode his pinto to the next camp to borrow from Jim Reardon, their outfit's foreman. There was a big float boulder in a buffalo wallow along his route, so I hid behind it and waited. As he rode past with his load, I jumped him. His horse shied, dumped him, and I choked out my money. He opened his eyes and gasped, "Hold it and I'll pay!" He produced a ten dollar bill and I said, "There isn't any change, just interest!"

Then I accused him of forcing me out since I'd refused to sell 'dips of snow' to the wagon skinners. All our men were dope fiends. They used a dried mixture of cocaine and boric acid, and a dip was all the powder they could scoop out of a cigar box on a dime. They would hill it on the backs of their hands and sniff it up their noses. The snow replaced their wages, and Baldy's camp was their major source of supply.

As the bookkeeper, he wanted me to sell the dope along with the other stuff in the commissary, so if anything happened, he would be in the clear. I refused. Instead, I went down the line to Jim Reardon's camp and told them what was going on. Reardon hired me to drive his grader, but after the first day, he said, "Tom Payne, you'd better beat it. Baldy has gone for the Mounties!" So I scrambled under the hay bales on a freight wagon to Bassano.

I wrote to Mont that I'd been to Saskatoon to apply to either the 105th Battalion, or the Motor Transport Division, but had been turned down again and given my exemption papers: "Unfit medically for any branch of the Services!" I told him that I'd send a duplicate copy, and insisted that if he ever heard of any curious persons wondering why I was not in khaki clothes to please "Call them down for me!"

ALICE PAYNE When World War One broke out in Europe, many British-born men hurried to join the army, especially those who'd settled on the prairies. For those who didn't pass the medical, the grief and embarrassment at being refused was not only public but personal. Dad was no exception and tried twice to offer his services. Although he was fit, strong, and smart he was finally rejected in 1917 because of an incipient hernia. Ironically, his life took on a pattern of working long hours at hard physical labour.

Stan McCormick For a while Tom worked as a pen-pusher and cost accountant on a Big Four Farm in Flaxcombe, Saskatchewan. This was one of several large farms run in the grand old style of American or British interests. It had ten or twelve big tractors and was owned by the Scottish Co-op Wholesale; they still have a seat on the Winnipeg Grain Exchange and a string of elevators. Most of those big grain farms got overbalanced and went broke.

Alice Payne Of course, Dad didn't know that yet. He'd been spending the winters trapping, and the summers farming, and was full of enthusiasm for both. He said so in a letter to Mont:

> Yes, those wolves were a fine bunch. I have made them into a robe and I may get some more next winter. If I do, I'll ship them over so you can have a decent robe for yourself. Farming here is sure booming. Ye gods there is a fortune staring a man in the face, if he only has the cash to get a good start. Land here can be bought for fifteen dollars an acre, as good as any land that Strutt has, and it's capable of producing up to fifty-five bushels per acre in a favorable season. With ordinary luck, it leaves quite a margin for one's pocket. Mont, you have a far scheming mind, is there any way I can get hold of a dollar or two from my old man? What if I got Strutt to write to him? After the war I'm coming home, and I'll bet you fifty to one that I'll get some money then, or there will be one awful smell of guts around our joint at home, either mine, or my old man's!

When Dad returned to England he found the weeds growing wild around the house, and his father an alcoholic. His medical practice had suffered, and he'd stopped operating The Retreat as a lunatic asylum. In fact, things were no better than when Dad had left years earlier.

He quickly decided to board with his brother's family. When Dad and Harry left home, Marian and Bill were young. Dad had never really known either of them. Marian had married John Taber, another Felsted boy, and they owned a grocery store in Witham. His brother Bill had become a doctor in the best family tradition. Bill impressed and amused Dad, and the two got on well together.

TOM PAYNE Bill was a skilled knife man, and was appointed House Surgeon at Guy's Hospital in London for three years. Afterwards he went to Colchester and had a tremendous practice for twenty years. Poor Bill! Drink got the best of him, and he died at the age of fifty-eight after spending his last years at the Marlborough Head Hotel. He loved his work and his patients, but was very lax at collecting fees or keeping accounts.

I remember being with him in London. Summoned to make a house call, he said, "Tom, you come along! You might learn something." It was a terrible night. I followed along until he found the address. It was a real pea-souper. We had to keep one hand on the wall or get lost.

Bill opened the door and we went upstairs to the bedroom. There was an old girl all propped up on pillows, lying back with her eyes closed. Her hair was all combed out, and she had a little rouge on her cheeks. Bill went over to the dresser, a high one with a little thin mirror along the top, and set down his doctor's bag. As he opened it, he saw in the reflection that she'd opened one eye to spy on him. By the time he turned around, she was lying with her eyes closed.

So he picked up the water jug on the dresser, and carried it over to the bedside. With his other hand, he pulled back the covers, and poured the whole contents on top of her! Then he closed his bag, said, "Good day madam!" and walked out. When we landed on the street he said, "Well, she won't call on me again!"

Bill's friend and mentor, Dr. Harrison, was the coroner and local bookie around Braintree. His drug dispenser ran accounts through the open window where you handed in your bets on horse races. The sky was the limit for a lot of wealthy families. Every Sunday, Harrison ate dinner with old man Taylor at the White Hart Hotel: roast beef and horseradish, liberally washed down with Bass's ale, followed by Stilton cheese.

One Sunday while I was there, an old woman with a shawl broke in, crying, and yelled, "Oh, Dr. Harrison! My husband fell downstairs and broke his neck!" Harrison said, "Go home my good woman. Dead people are no use to me!"

His morning surgery was like a pantomime. It was held in an oak-panelled room lined with wooden benches. The latest racing tipster's sheets were strewn on a long table in the center. Harrison, holding a scotch and soda, would appear in the doorway and survey the crowd of patients in the room, nearly all of who were women.

"Will all good ladies with stiff backs please stand up?" Seventy-five per cent of all women have stiff backs, and most stood up. Then Harrison would turn to his assistant: "A bottle of number nine for these good ladies," he would say. She handed out the bottles, while Harrison retired to his den, had another drink, then tended to the balance of his patients.

His den had a desk, a big leather chair, and various trophies from all of his inquests: ropes, guns, knives or whatever was used. What with ferret collars, fishing rods, and other stuff, there was standing room only. On one of my visits, he plunged his hand down among the artifacts and produced a fine razor: "Here's a present for you. This is what Colonel Arkwright cut his throat with last week." I used that same straight edge for years until it finally wore out.

Alice Payne Dad was in an awkward situation as he'd hoped to earn some money in England. But he'd come home at a time when even wartime heroes had trouble finding jobs. The months passed, and his resources dwindled. He didn't even qualify for unemployment relief. It was a desperate time.

Dick Smith It wasn't until 1921 that I saw Tom again. He was broke as usual, and staying at one of Mont Everard's houses. Tom would often accompany me around the farms. One day, he got out an old gun and said, "Come on, Richard, take me to see my old man. If he doesn't dub up, I'll shoot the old bastard!" I was highly amused by this, and in a curious way, was all for it. The fact that the gun wasn't loaded hardly seemed to matter, and we stopped along the way to fortify ourselves at the local pub.

We came to his father's house in Witham and walked in. Tom laid his gun on the table and said, "Get out a bottle of champagne!" After we drank it, he said, "Well, what's it to be, cash or a bullet?" This was all bluff of course, but it worked! Old Dr. Payne forked over £440, which surprised Tom, and later, everyone else. After congratulating ourselves, we rolled off home. It would be the last time he would see his father.

Jim Pawsey I had some hope when I learned that Frank Payne had given up all his drinks and cigars. His son, Bill, might have been able to do likewise, but he was his own worst enemy.

Alice Payne Dad had sufficient money to return to Canada in 1923, and got hired onto another cooperative farm at Hughton, Saskatchewan. He'd been asked by Lord Dewar to check on it since he had invested heavily. It was a big operation and the farm people came to meet Dad in Winnipeg. He was all dressed-up in a bowler hat, and a fine pinstripe suit, looking very much the London businessman. Of course they knew why he was there, and they didn't want him around very long.

Stan Graber At Hughton, Tom was the foreman and I was the accountant. He was already a whiz on big tractors and machinery. It was obvious that it had taken many years of practical training for him to acquire enough knowledge about the workings of the internal combustion engine. He stayed for the whole summer.

Tom drove a long-hooded, straight eight Studebaker. The motor, between the floorboards and the front axle, was designed to produce more brake power than one of our 30-60 tractors. It had a 6-volt battery and generator system, which permitted a self-starter and electric lights. This giant of transportation looked like the side of a slow freight going through Hughton as Tom cruised the gravelled roads of the farmyard.

He was a big man. No matter if he dressed in his Sunday best tweed suit, with no hat and tousled hair, or in his greasy everyday work overalls, his presence was felt. He stood out in a crowd and would not tolerate laziness or inefficiency in others. Once, when guiding a high ranking Royal Canadian Mounted Police (RCMP) officer on a rescue mission in the far north, Tom became impatient with the slow progress the officer was making in paddling the canoe. So he said bluntly, "Son, give me that paddle. We've got to get going and I'll show you how to do it."

In 1926, I was working in Saskatoon for the distributors of Hudson and Essex automobiles for northern Saskatchewan. Busy at my desk I looked up, and there stood Tom Payne, smiling broadly. He was on his way to Edmonton more for a change of scenery than for any other reason he could think of.

BERT WHEATLEY Tom worked sporadically for the municipality of Winslow, and then went threshing for M.C. McCormick at Dodsland around 1928. The town had raised some money for road improvements and we got the contract.

There were three of us on the crew, all Englishmen. Tom was the engine man and tractor driver. I was cook and handyman, and the other fellow manoeuvred the grader. We had quite an outfit. For the first two years, we used an Aultman-Taylor tractor. Top speed was two miles per hour. The back wheels were seven feet high, and Tom worked hard to run it; just steering it was tough.

Later, we used a Caterpillar 60, a small tractor, which was easier to operate. I was making about four dollars a day, and I think Tom was making ten. Considered a very good wage. Of course, we worked hard, spending about ten hours a day on the roadwork, and most nights fixing the equipment.

We lived in a bunkhouse on wheels, hitched to the tail of the tractor. Often after supper, we would drive into town for a beer at the bootleggers. Having sharpened our steel at the blacksmiths, we would buy groceries and meat. If there was time, we'd spend the rest of the evening with friends. The twenty-mile trip was worth it because Tom always needed tobacco. It was pretty expensive stuff. After he caught everyone smoking his cigarettes, he bought a pipe and filled it with an imitation kinnikinick, broken-up shag tobacco dipped in wine!

One night, we drove into town in Tom's Ford coupe, a nice little car with a canvas top. I went to Druid, and he went on to Dodsland with the grader-man to get some machinery at the blacksmith shop. A sudden storm blew up with rain, thunder, lightning, and tornado-like winds. When Tom came back to pick me up, his car was covered with mud. The dirt roads had no gravel yet, and if you got stuck, you sank in axle deep.

Within a half mile of camp, we ran into an old farmer who said, "I don't think there's any use you boys going home. I don't think your bunkhouse is there." We went anyway, and found it had been demolished. Apparently, it had moved a little ways, and then tripped over a boulder and upended. The wheels got mired, and the walls went in four different directions. The roof landed thirty feet away. About twenty cases of empty beer bottles slid out, but the lamp was still sitting there with the glass intact, and the clock was still ticking! The only thing we could do was gather up the pieces and find a hotel.

Tom could really drink beer. During his last winter here, a bunch of us batched together in the bootleggers shack. On one occasion, Tom came in after work and found a group of admirers waiting for him. One man stood a case of Molson's beer at his feet and said, "Tom, you can have for free all you can drink in twenty minutes!" I guess they had a bet on, and he was a big guy. Well, he started to drink. Out of twenty-four bottles, I think there were only seven left. He didn't get drunk, but that was the one and only time he ever asked me to drive his coupe.

BIG FARMS IN EARLY SASKATCHEWAN

**One special Scottish farm worker was Tom Payne**

**A group of English and Scottish lads had come over to the Big Farm for the 1923 farming season. The older man, second from the right, is the Big Farm foreman, Tom Payne, who was the first to strike big gold in Yellowknife.**

*Grainews, June 1993. (Courtesy of Stan Graber)*

Stan McCormick Tom and my Dad, M.C. McCormick, got on well, and every fall, when the road grading crews shut down for the threshing season, he came to work for us. It was up to him to keep the threshing machinery going. The big separator and the engine were both tethered to a long belt. Compared to a tractor, it was very tricky. You had to keep an eye on it. Everything depended on those pieces running smoothly and continuously.

Tom was conscientious, and only once didn't show up on time. That day, the old man and I went out to the field, and got the motors going. Along came Tom. He was only about three-quarters of an hour late, but mad at himself and everybody else. No one said anything. Tom was a damn good man. That night he ran the Ford into the garage and didn't take it out again until threshing was over.

'Tommy' Payne Dad told me about a steam tractor that he used to operate in Saskatchewan. It was a 1906 Case, and fairly large. He and an engineer were running it over a hill, and apparently the water tank got low. The heat built up on the crown sheet and the soft lead plug, a safety device in the firebox that kept everything wet, burned and melted out. There was nothing else to do but remove the coal, shovel everything out on the ground, and crawl in to replace the plug.

Years later, when he and I were out driving southeast of Edmonton, we came around a corner to find one of those big steam tractors sitting on the road. It had completely run short of water. Along came the farmer in a pick-up truck with a couple of tanks, and Dad helped him get the pumps working, and get her all straightened out. Then he cranked this old tractor around, zapped her off into the guy's field, and helped turn over a few acres. I was really impressed. I didn't know he could do things like that.

Stan McCormick It wasn't all work either. We had some good times too. There was an old bachelor, Rudolph Myers, who had worked with my father in the States. Myers had a piece of land a half-mile north of ours. Every night he used to walk to Dodsland along the railroad tracks, which went right past his house. One Halloween night a group of men, not kids, grown men, decided to go to Rudolph's house and play a trick on him.

Eight of us piled into two cars. But when we got there, two of the fellows sneaked off across the road. Meanwhile everyone else was busy taking a wagon apart and reassembling it on top of the roof! About the time that these two characters figured the job was finished, they fired off a couple of shots over everyone's head and scared the living daylights out of us. We were off that roof in no time, racing for the cars.

But Tom's wouldn't start. He tried to crank it and it wouldn't go. You could hear him yell a mile away, "For Christ's sake, choke it, choke it!" We drove off, then stopped to count noses to see if anyone had been shot. Two guys were missing, so we went back. There they were, rolling on the grass laughing. We had as much fun over the double cross as we did thinking about Rudolph coming home.

That night, Tom and Rudolph came over to our house for dinner. Rudolph was still pretty sour after discovering the prank the next morning. Afterwards, he said, "McCormick, how about you and the fellows coming to help me with a little job I have to do?" Of course, Tom and I knew what it was, but my father didn't. There was plenty of daylight left, so we walked the three-quarters of a mile over there.

*Stan McCormick's threshing outfit, Dodsland, Saskatchewan. Photo by S. McCormick*

It wasn't until we were in the yard that my father saw the wagon on the roof. Did he laugh and laugh. Of course that made Rudolph even madder. We took it down, and no one said much until we were halfway home. Dad said, "I wonder how the hell that old wagon got up there?" Silence. Then he said, "I'll bet you two bastards did it!" It rather tickled him. And the funny thing was, the two who had fired the shotgun were across the road watching us dismantle it.

Sometimes in rainy weather, Tom and Bert visited Agnes and me to talk or play cards. There weren't many married couples, and all the bachelors enjoyed our hospitality. In the evenings, he would entertain us with stories of other people he'd met. He once referred to most single women around here as being poor specimens of humanity; dead from the toes up.

One woman that he'd boarded with was an artist at cutting meat. She would cut a slice of beef so big that it would cover your whole plate, and heap lots of potatoes on top. But when you got down to eating the meat, it was so thin that you could fold the whole slice up and eat it in one mouthful. If you held it up you could read the newspaper through it. And all the time she was slicing, she would be making comments about how generous her servings were.

Another gave Tom extra chores to do in the evenings. Every night after supper she would ask him to see if the chicken house door was shut. He would go and look, then come back in and report. If he said the door was open, she would cock her head to one side and say, "Well, it might rain tonight, I think you'd better go out there and close it." But if he said it was already shut, the reply would be, "Well it needs some fresh air in there. I think you'd better go out and open it."

So every night Tom made two trips out and he got tired of it. One night he came in and when she asked about the chicken house door, he went out and closed it. Then he came back and said it was open. But after the ritual, instead of going out a second time, he just sat still, and after awhile said, "It is shut."

ALICE PAYNE Dad decided to take his own advice and try his luck at prairie farming with the help of his father's £440. He broke a section of land near Hoosier, Saskatchewan, and put it into flax. Most of the crop froze. After harvesting the rest, he sold the farm and paid all his expenses. He had just enough left for a case of Dewar's Ne Plus Ultra whisky. He took a bottle to town, everyone got tight, and that was that. Dad decided he wasn't going to be a farmer after all, and set his eyes on the north.

# Hudson Bay

## Chapter 3

Alice Payne For most bachelors, seasonal work on farms meant that winters were spent elsewhere. The country north of The Pas was the natural place to go. Prospectors had started out looking for gold, thanks to a Geological Survey of Canada report in 1896. Instead, they found the large base metal deposit at Flin Flon, which remained unmineable until the engineers figured out how to refine the ore. Construction started in 1927, with a mill, copper smelter, zinc refinery, and the necessary infrastructure. About a hundred miles beyond there, the Sherridon copper-zinc mine, discovered in 1922, was close enough to the end-of-steel to be economical. All this activity spurred more development.

Gordon Willsie I went to The Pas in the spring of 1923 and then prospected the country around Flin Flon and Herb Lake. Conditions at The Pas were very lively with Indian dances and considerable frolicking. For five years I was the official timekeeper of the famous 200 mile Dog Derby races. My father-in-law, Louis Allard, the teamster, worked for Tom hauling freight. He was well known up there during the twenties.

Jack Moar When the preliminary work was being done at Flin Flon, Hudson Bay Mining and Smelting development needed a man to operate a shovel. Tom could run both steam and gas powered engines, and was recommended by the contractor to scrape overburden from the future mine site. Along with the muskeg, he dug a pit out of an ore body. Every shovel bite was worth over a thousand dollars because of the gold content. This is what really inspired his interest in mining.

All freight hauling was done with sleighs on winter ice roads. Men on snowshoes with dog teams began construction in the fall. When the trail was ready, horses and sleighs pulled water tanks over it, gradually building it up enough for the tractors to support heavier freight. They constructed an ice road from The Pas to Flin Flon some thirty feet wide and three feet thick, complete with marshalling yards, turnouts, halfway houses, and electric lights.

Linn tractors were the main pieces of equipment. They looked something like a truck on skis or a skidoo, with tractor treads behind, and could pull more than they carried. Tom signed on as a driver and worked for Harry MacLean, the flamboyant contractor who pioneered the Linns in northern exploration. MacLean later became famous for throwing money out of Toronto hotel windows.

Tom Payne At Island Falls, I met an old Russian living on the Churchill River who made lovely caviar. He was a real expert. He'd once made it for the Czar and knew how to pickle-smoke the sturgeon roe. His only trouble was a lack of containers. So I gave him six empty forty-five gallon oak wood barrels plus a quarter of company beef. In return, I got a barrel of caviar for R.E. Phelan, the President, who wanted to pay me for it. But I had to confess how I got it, and he got a big kick out of it!

Alice Payne The hydroelectric plant at Island Falls required sixty miles of transmission lines, and a branch line connected the minesite to the railway. The freight haul to the site was a big job: 23,000 tons of heavy materials had to be transported over sixty-nine miles of terrain before winter's end. One forty-hour, nonstop, round trip consisted of eleven Linn tractors, each pulling sleds. There were five road gangs along the way, and strategically placed water trains to ice the portage and keep the roadbed smooth. The record for one trip was 112 1/2 tons, although the average was eighty-five.

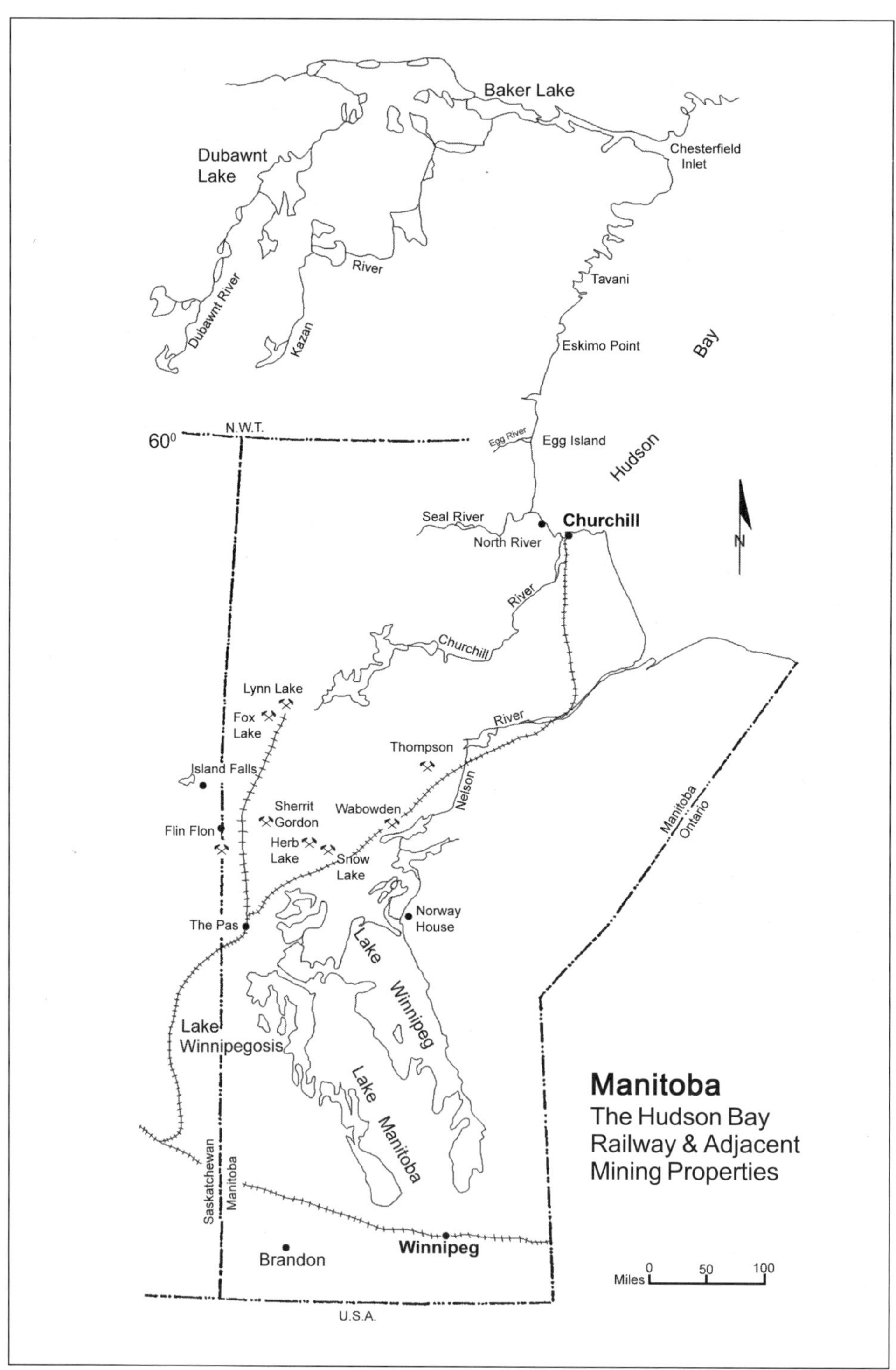

Baker Lake
Dubawnt Lake
Chesterfield Inlet
River
Tavani
Dubawnt River
Kazan
Eskimo Point
Bay
N.W.T.
60⁰
Egg River
Egg Island
Hudson
Seal River
Churchill
North River
N
River
Churchill
Lynn Lake
Fox Lake
River
Thompson
Island Falls
Sherrit Gordon
Wabowden
Nelson
Manitoba
Ontario
Flin Flon
Herb Lake
Snow Lake
Norway House
The Pas
Lake Winnipeg
Lake Winnipegosis
Lake Manitoba
Manitoba
The Hudson Bay Railway & Adjacent Mining Properties
Saskatchewan
Manitoba
Brandon
Winnipeg
Miles 0 50 100
U.S.A.

Typically, one load consisted of cement, contractor's equipment, extra gas, wire and steel, lumber, and 1,000 tons of provisions to feed 800 men for one year. In other words, 450,000 pounds of beef, and 15,000 eggs, or half an egg per man per day! An ammonia plant kept the food frozen, and the power plant ran the camp. All this material was hauled in during the winter of 1929. Then the tractor crews were cut loose, which freed Tom for his next adventure.

Thayer Lindsley, an American who had brought Sherritt Gordon Mines Limited into production, sent his prospecting organization, Dominion Explorers, into the far north on an imaginative task. Led by Lieut.-Col. C.D.H. MacAlpine, a Toronto mining man, they expected to move men and supplies to a base at Tavani, an Inuit village 300 miles north of Churchill, and from there, conduct detailed work around Rankin and Chesterfield Inlets, Baker Lake and the Dubawnt River.

To quote the *Northern Miner*, they were to "Seek out and coordinate the coming season's work of a score of prospecting parties, to ascertain winter conditions, to exploit the company's advantage of having wintered on the ground, and to generally review the far-flung northern prospecting situation." A caterpillar tractor train carrying supplies was sent from Flin Flon, and a two-plane scouting expedition was sent out on the first of two trips.

*Hudson Bay Mining and Smelting Company's test mill at Flin Flon, MB. G. Willsie collection.*

On March 25, 1929, pilots Stan McMillan of Edmonton and Charles Sutton of Toronto left Ottawa for Churchill with two Fairchild aircraft. They carried 1600 pounds each, including food, tents, engine covers and heaters, snow knives and emergency equipment. On board were Lieut.-Col MacAlpine, Norman Pearce of the Northern Miner, a geologist, a New York cameraman, a mechanic, a wireless expert, and a troubleshooter.

The well-publicized flights, 4000 miles for one plane, and 6000 miles for the other, turned out to be the longest winter air trips ever made in Canada at that time. After reaching Chesterfield Inlet and Baker Lake, they returned home by the same route, and flew over the two tractor trains on which Dad was crawling up the coast of Hudson Bay.

Victor Stevens I was introduced to Tom at Tavani, which means in Inuktitut, 'up there in the north.' I was with Dominion Explorers and they conceived the idea of freighting us our supplies by tractor.

*A Linn tractor hauling a load of winter freight to the Sheritt Gordon Mine. G. Willsie collection.*

Unfortunately there was a heavy flow of water on the sea ice, covered by a thin ice coating. The tractors got bogged down and were unable to proceed. Tom was one of the drivers.

TOM PAYNE    Wages were $500 per month including board, which was good pay in those days! In order to qualify for the job, every Cat skinner had to be capable of taking a Linn Cat apart and if necessary, rebuilding it. I set out with Harry MacLean's outfit from Flin Flon on the Hudson Bay Railway, to the end-of-steel at Nelson River. It would take four years for workers to extend the line right to the coast.

Along the way, the Russian and European immigrants rode in the boxcars. How they sang! Every verse ended with "God bless the CPR." For many, that was the only English they knew. The train would slow down somewhere in the barren lands, and a man would yell for six guys to get off at intervals of about 500 feet. Any six, mind you.

They would jump into the snow with their bundles of clothes and grub, followed by stoves, wheelbarrows, picks and shovels.

*Dinner served on a shovel. G. Willsie collection.*

By the next day they would have built a shelter out of muskeg, snow and mud, and baked the loveliest buns you ever tasted. Soon they'd be working, digging puddle ditches alongside the track. All for seventeen and a half cents an hour, and often in forty to fifty degrees below with a bad wind. It made me wonder about the conditions they'd left behind in their own countries.

From Nelson, we assembled the sleighs, loaded the freight, and set out for Churchill and Chesterfield Inlet. The trees thinned out and became scrubby, with their tops just poking through the snow under our treads. No wonder the Eskimos called it 'the land of little sticks.' There was no protection against the wind, and we froze and suffered terribly until we got some fur-lined clothes. The main thing was to get up right away in the morning and shake the frost off your sleeping bag before it melted. If you got it all off, then you were alright. If not, then you iced up.

On our arrival at Churchill, the military commander forbade us to enter inside the twenty-mile square, single wire fence that surrounded the place. We had one hell of a row over it. I asked how far and how tough it was to get around that damn palisade. Since we had come 250 miles, a few more steps made little difference.

We were just getting ready to detour around when a dog team arrived with a telegram from Ottawa: PLEASE RENDER THE

*A railroad bridge under construction. G. Willsie collection.*

MACALPINE EXPEDITION EVERY ASSISTANCE. In we went. We were put up in the staff house, and fed and liquored in the officer's mess. The army Cats were pulled out of their garages and parked in snow banks, while ours were greased and serviced. Heaven! Then Sergeant S.G. Clay took charge of our outfit, aided by three Eskimo guides.

Clay was a fine character and could speak fluent Inuktitut. Five years earlier, he had lost his wife in a tragic sled dog attack while on a long patrol to the Thelon River. She was a good dog skinner and often visited Eskimo settlements by herself. People believed that one of the dogs snapped at her in play, drew blood, and then the whole pack set upon her. They were driven off, but it was too late. There was no doctor, so the priest, a constable, and a fur trader amputated her leg. But shock overtook her, and she died.

We were all set to leave for Chesterfield Inlet when a wire came from MacAlpine: UNDER NO CIRCUMSTANCE LEAVE CHURCHILL FOR THREE WEEKS. THE EQUINOCTIAL STORMS ARE DUE. Ironically, those three weeks were the mildest and loveliest weather of the entire winter. It was clear and cold, the moon shone, and the huskies howled. When the wire finally came to pull out at once, it was midnight with a howling blizzard and sixty degrees below!

*Bogged down in the ice of Hudson Bay. G. Willsie collection.*

Clay asked me what to do and I said, "Pull out! We will either sink this outfit or scatter it along the ice."

We crossed Button Bay and travelled in a heavy ground fog to a settlement called North River. There we got stuck on an overflow, caused when thin layers of ice are covered by tide and river water. I was off-shift and sleeping in the rear bunk when the first Cat partially sank in the delta. The operator jumped from the cab and yelled "Get the hell out of here!" The second Cat was trailing us by a mile. You could just see its headlights.

I pulled on my *kamiks* (sealskin boots) and parka, and found myself standing in knee-deep water. The slick ice was tilting towards the hole where the Cat was still sinking, and I started to slip backwards into the larger pool surrounding our outfit. The mixture of fog and icy air was torture. Finally, I hit a dry section and my rawhide boots froze to it like a sheet of iron. I yanked them off, and wearing just my socks, ran back to warn the other tractor driver, who was towing the grub and cook shack and two sleighs, each with ten tons of freight.

When daylight came, we discovered that it wasn't as bad as we first thought. The overflow was only about three feet thick and we had

*Making an igloo near Churchill, 1929. G. Willsie collection.*

a long three-quarter inch cable, which we hooked on to the second Cat. We had picked up a few railway ties at the Churchill army camp in case they came in handy, and they sure did. We laid some in front of the engine for a back lift, and winched the Cat out with one enormous heave.

Then we turned north toward Egg Island. But Nikvik, the Eskimo medicine chief for Hudson Bay, took exception to our stopping there. We spotted him and his son, Stockings, on the ice crawling on their bellies. They had left their *komatik* (sled), and the dogs, and were pushing their rifles ahead of them. I couldn't see very well through the fog, and I thought they were stalking a polar bear one foot at a time.

Back came Stockings shaking his head and holding up his hands. He was in one hell of a shape, and spoke a fair broken English. "Noak," he said, "No, evil spirits." We suggested they camp with us, in the lee of the islands, as they were all pretty exhausted. So they prodded around everywhere with their Noak sticks trying to find good snow with which to build their igloo for the night.

This procedure was always the same: The Eskimo stop to camp at a place where they know the snow is opaque and solid, not sugar snow, and leave you there. Then they go and snipe-bill around in about a half-mile circle endeavoring to find a better spot than where

*Four tractor drivers, Tom Payne on the right. G. Willsie collection.*

you sit with the sleigh and dogs. During this period, you ice up, and nine times out of ten, they can't find a better spot!

It took us over a month to get to Seal River, less than fifty miles from Churchill. There is a great granite hump in the river delta, dotted by small clumps of willow bushes and numerous channels. The landmark can be seen from twenty miles away. Here we got into trouble, proper trouble. We got caught in a blizzard with hard driving snow, and it was bitterly cold.

We kept going until Stockings and Nikvik said to camp, but in the meantime, one Cat unhooked from the loads and went through. Only the tip of its snowplough showed through the ice. I came out through the top of the cab, which was of soapbox design: if you sank, the lid floated off and you followed it up to the surface.

We freed our second Cat from its load and tried to rescue the first one, but couldn't budge it. So we returned to the cook shack and bunkhouse, and drained the cylinder oil out of the engine. Before shutting down, we ran it for five minutes with coal oil so that we could crank start it later. This involved warming it up with five men pulling on a rope hooked to the handle. Then damn if that Cat didn't start to sink too, right beside the cookhouse!

Fortunately it settled on a ten-inch layer of ice with another two-foot layer below. There was no water between the layers as the tide was out. But the next day, the eighteen inch surge through the hole where the first Cat lay reminded me of a frozen geyser. The second Cat stayed high and dry, thank God.

Things were getting grim. The reporter, who was supposed to be covering our progress, began to get loopy. All he would do was sit, hold his head, and groan. This went on for a few days, and then his whole system quit! He wouldn't eat and his bowels failed. I was at my wit's end trying to help him, and sympathy hadn't worked in the least. Finally I told him to quit worrying about his wife as someone else probably had her out tonight. This snapped him out of it somewhat, and I made him drink a big draft of black strap molasses. He improved a little, but was still a concern.

At this point Stan McMillan flew over with a gunnysack of mail from Churchill. He dropped it as he couldn't land on the overflow, and it contained unsettling news: a Major Baker was coming up to relieve Sergeant Clay, accompanied by four taxi drivers from Winnipeg to take over our jobs. According to Baker, our outfit had apparently hired useless Cat skinners, and he was going to make four or five trips to Chesterfield Inlet and back in no time flat.

Major Baker's fame preceded him. Apparently he'd had a lot of experience in Africa and Egypt freighting with elephants and camels. He arrived with two Eskimo guides and a dog team, drunk as a lord. "I am Baker, Major Baker!" he announced. "Well, how de do, Baker," I said. "What's your name?" he asked, and I replied, "Payne, Tom Payne," thinking I might as well have a handle to my name too. With that, he retreated into the cook shack and took another snort, then re-emerged and said, "Payne, bring in my bed roll." This was too much for me and I told him, "What's the matter with you carrying it? You are a big strapping fellow!" So we didn't see eye to eye from then on.

Next morning Baker turned up, all spit and polish, and said to me, "Payne, this is one beastly mess!" So I repeated, "One hell of a mess." Then he asked how I proposed to get this machinery out of the ice. He had never seen a Cat before and didn't know how to use a jack. I couldn't stand it any longer. "We've had advance notice of your arrival and understand that you're taking over. So go ahead. There are eight or ten damn good men here, and they will do everything you tell them!" Of course he didn't have the expertise to deal with the machinery or the situation. He just stood there looking angry.

Thankfully we continued under Clay's direction, since Baker showed no signs of sobering up. With time and a lot of sweat, we got the stranded Cat free with chain blocks, large railway jacks, and chains, inch by inch. I thought we'd never make it. To lighten the load on the bunkhouse flat, we chopped the boxes of freight loose and heaved them onto the ice.

One box marked FLIT broke open on impact. The stuff didn't look like the well known fly and mosquito killer, and someone was bold enough to stick their finger in it. Low and behold! It was good old navy issue rum. Twenty-four, five-gallon cases! Needless to say we decked it in short order; an overdue reward for our heroic efforts.

Meanwhile, Baker had learned from Stockings that the ice inside the tidal crack was piled mountains high, all the way out to the hinge ice. So he ordered us to travel to Chesterfield on young ice! Clay, Baker, myself, and the other Cat skinners held an urgent council meeting. We informed Baker that his place was in Africa in a tent with a swagger stick under his arm, and insisted that Sergeant Clay take over and lead our outfit back to Churchill.

Our old track had drifted in, but we made it back in good time since we knew what pitfalls to avoid, having explored them so well with sinkings on the way north! We got our paychecks and disbanded, glad to kiss that mission goodbye.

Baker travelled back to Churchill via his Eskimo dog team. By spring he had returned in the same aircraft with Lieut.-Col. MacAlpine and company. They got royally lost that time. Dominion Explorers spent a million dollars trying to find them and nearly went broke. The Inuit eventually escorted them back to a Hudson's Bay Company trading post. They had put in a tough time but had survived. Norman Pearce, so legend had it, had to trade his gold watch for a few meager fish.

After that, my Métis friend, Andy Beaulne, and I went trapping on the Seal River. We lived in a combination sandbank and igloo. Our house was in a sand bar depression with a front of snow blocks and driftwood from the delta. When we set up at Seal River, we shot lots of white whales for fox bait, and beached them for a mile along the frozen shoreline. Snowdrifts covered them up. We hoped for caribou to wander in, bringing the wolves with them. The foxes would feed off what the wolves left behind.

At first, there were no Arctic foxes. Then we awoke one morning and you couldn't drop a dime in the snow without it landing on a fox track. There were millions! We shot and trapped to our heart's delight and piled them around the igloo. Every trap had a fox in it, and every buried whale had a bevy of foxes gorging on it. You could hear them feeding under the snowdrift. The drift had a hole at each end. I would poke in one hole with a bit of old wood and Andy would club the foxes on the head when they emerged at the other. What a slaughter!

We threw the fox pelts in a heap on the komatik, and stored them beside the igloo.

The big problem was thawing and skinning the carcasses after things settled down about ten days later. I was hoping the foxes would linger awhile to clean up the white whales, but they left in one night, and there was no sign of them the next day. A good thing, really, because we were late drying and baling our skins. We had one hell of a valuable catch, and break-up was coming.

By this time, seals were getting their push-ups through the delta, which look like muskrat mounds on a lake. What you do is locate a good strong one, stand downwind on a caribou skin, coil your fifty-foot sinew rope very carefully, get a willow stick, and spear into the center of the push-up. The stick should be just above the water line. Then you pack snow all around it, and turn all the dogs loose to visit the neighbouring ones.

The quiet time is when you sit waiting and watching, spear in hand. Soon you hear a glob-glob-snort as Mr. Seal comes out to sit on his ledge. You know where his nose is, as the stick poking out of the push-up begins to move. Then you drive in the spear, lashed with a string of sinew rope, and hang on tight.

The shaft is very precious as there is no timber, and you mustn't lose or break it. Mr. Seal then takes his dive, and it takes one hell of a strong man to survive its first plunge. You have to watch the rope, and not get it tangled around your leg or it would break in a second. In the worst case, the rope could drag you under. When the seal plays out, you chop all around the hole. Then all the damn dogs roar in and chew flaps out of Mr. Seal, which helps drag him up on the ice.

If your dogs are really hungry, you can't save your seal. What's left, out of a mad scramble, is a bloodstained ice patch! The smell of blood drives them mad. All the savagery of the wolf comes out. But since there are thousands of seals, you soon get another one and your dogs, gorged with blubber and blood, are not so dangerous.

Once Stockings and I went hunting seals near Chesterfield Inlet. We were camping right on the ice: dogs, sleighs, knives, everything. It was no use being on land because the seals were too far out, and the shore ice is too thick. The seals can keep one air hole going all winter, and you could see the dark water through it. We just sat there until Mr. Seal surfaced for a snort of air, then nailed him with the harpoon.

Once we'd trapped all the seals in an area, we loaded up our komatiks and moved on to a new spot. We were doing pretty well, in spite of the odd blizzard, but there were lots of seals. We'd kill a bunch, skin them, then go around and pick up several hides and whatever meat we needed. We had the best fed dogs in the world!

When it was my turn to do the rounds, a bad storm blew up and we got separated. Stockings was my best compass, so I decided to sit it out. I turned over the sled, wrapped a few sealskins around me and the dogs, and went to sleep thinking about him. He was one hell of a good guide. In the middle of a whiteout or storm, he'd go into a trance, shut his eyes, turn around a few times, and then say, "That way." He was never wrong. I just followed after him and sooner or later we would hit camp.

When I awoke in the morning, it was a lovely clear day. All I could see between me and home was water! I was on the edge of an ice floe. There was no sign of Stockings or our base camp anywhere. So I sat down, melted snow on the primus stove, and thought how lucky I was that the dogs were fat! I had one whole seal, plenty of skins, and my tea and sugar. On the sled were my emergency fish net and my gun. Not bad.

As my ice floe drifted, I considered the possibility of running into a boat in the middle of the bay, that is if the last one hadn't left for the season. I emptied the komatik, turned it over, and made a little pile of kindling to put underneath for a signal fire. The next job was to build a snow house so I wouldn't get devoured by the dogs along with the food.

Then I settled down to wait. I caught a few fish and spent my time gazing at the horizon. On clear days, when you could see the sun through the clouds, I tried to figure out my location. I couldn't decide whether I was out in the middle of the bay going towards Ungava, or if I was drifting in circles. But no boat came and no land appeared. Then one day I crawled out of the igloo to see a big crack right through the middle of my ice floe, and salt water only fifty yards away. Fortunately, it didn't spread any further.

When I finished up the sealmeat, I started eating the fish. I fed a few to the dogs, so they wouldn't eat me. I never liked travelling with dogs; it's hard enough to keep from starving to death without worrying about them. In the end, I had to eat some of them too.

Eventually, my floating camp stopped moving. I asked myself, "What would Stockings do?" So I hitched up what remained of the dog team, headed east, and hoped I'd made the right choice. Sure enough, my luck held. I had landed on the eastern side of Hudson Bay. I was never so glad to be back on land! My stove had run out of fuel, so my first priority was to make a fire to thaw out. Then I set up my camp and got ready to stay the winter. Rabbit stew tasted wonderful after all the raw fish. Just in case anyone appeared, I made a signal pile out of dead spruce trees.

The sight of a boat approaching sent me scurrying to ignite one hell of a smudge fire. The sailors picked me up with all my furs and dogs. No one would believe the story, but neither could they explain what on earth I was doing there otherwise. I had a hot bath and a good meal, and landed up in St. John's, Newfoundland. All I had was my sealskins and fur suit. I sold the skins but kept the suit just in case I needed it again.

Interesting business, hunting. Here's the point where evolution comes in. Lots of caribou equals lots of wolves equals lots of foxes. Then the wolves get fat and kill off the caribou and the herds diminish. So do the wolves and foxes. They seem to move in five-year cycles. Andy Beaulne returned to Seal River the next year and caught just one fox. Well, 'nuff said about that.

# CHAPTER 4

FRANK GUMMERSON    Tom and I were about twenty years apart in age, so we had a casual acquaintance. But I did run off a copy of his employment record in the local personnel office at Hudson Bay Mining and Smelting here in Flin Flon. He was hired in 1929 and left in June 1931. The coded card meant 'cut in workforce,' and a great many men were laid off at the same time. Tom's freight hauling job had pretty well ended anyway. A couple of other men knew him on the same casual footing as my own. It stuck in Harry Guymer's mind that Tom intended to try prospecting in the Yellowknife country after he left here.

ALICE PAYNE    Fortunately for Dad, aerial prospecting had opened up the far reaches of the western Arctic for exploration, and the eastern shoreline of Great Bear Lake was the scene of the next boom. I spoke with a Swedish fellow who had been out in the bush with Dad, and when I asked him what equipment he took along, he replied, "Oh, nothing: ve just vent!" and that's just the way it was.

In 1928, Colonel MacAlpine, the original enthusiast, discovered silver near the mouth of the Camsell River and established the White Eagle Silver Mines. But most of the action began when Gilbert Labine peered out the window of Punch Dickins' airplane and spotted some discoloured rocks. They were stained heavily with reddish hues, what

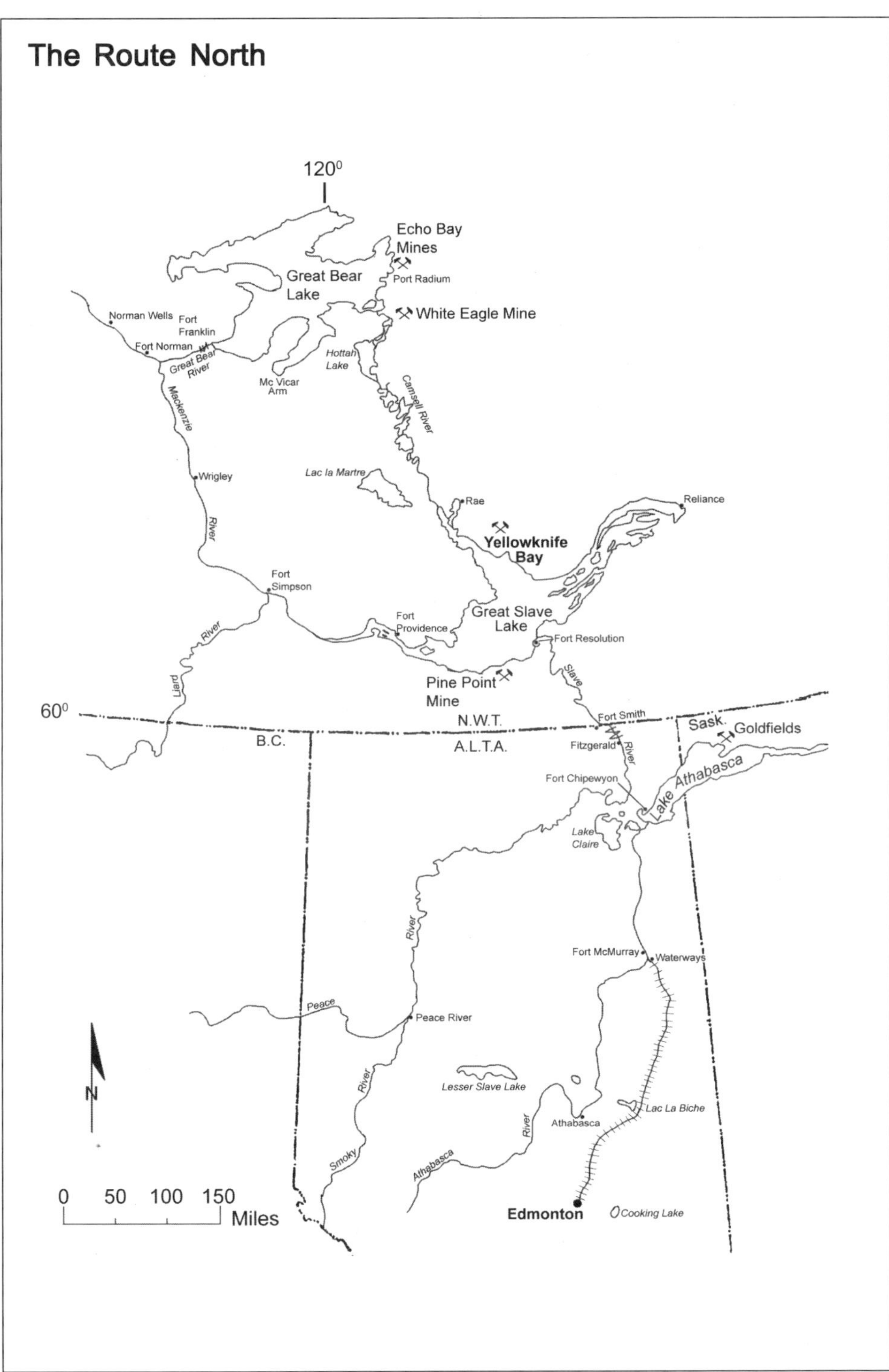
The Route North
120°
Echo Bay Mines
Port Radium
Great Bear Lake
White Eagle Mine
Norman Wells
Fort Franklin
Fort Norman
Great Bear River
Hottah Lake
Mc Vicar Arm
Camsell River
Mackenzie
Wrigley
Lac la Martre
Rae
Reliance
Yellowknife Bay
River
Fort Simpson
Fort Providence
Great Slave Lake
Fort Resolution
River
Liard
Slave
Pine Point Mine
60°
N.W.T.
Fort Smith
B.C.
A.L.T.A.
Fitzgerald
River
Sask.
Goldfields
Fort Chipewyon
Lake Athabasca
Lake Claire
River
Fort McMurray
Waterways
Peace
Peace River
River
Lesser Slave Lake
N
Lac La Biche
River
Athabasca
Smoky
Athabasca
0 50 100 150
Miles
Edmonton
Cooking Lake

the prospectors called a gossan. Pinks, blues, and greens were also visible, and the outcrop appeared to contain many veins and faults. This complex display reminded Labine of the rocks of the Cobalt mining district. There, the ores of nickel, cobalt and silver oxidized to a variety of colours, in particular the pink 'cobalt bloom.'

Knowing that the Geological Survey of Canada had reported copper and cobalt in the area in 1899, Labine sent samples to Ottawa for testing. Sure enough, the nickel-cobalt-silver mineral suite was present, but there was a bonus: a uranium ore called pitchblende. This in turn contained radium, used to treat cancer. Radium was worth $50,000 or $70,000 per gram since at the time, it was only being mined in the Belgian Congo. Later, when radium became less important, Labine's Eldorado mine produced uranium for the Manhattan Project.

When news of the find leaked out, everyone rushed up to Great Bear Lake. All the major mining companies or syndicates had staking parties in the field. The Geological Survey of Canada urged their scientists to use aeroplanes to assist in mapping, but D.F. Kidd, who lost part of his hand due to a propeller accident, conducted many of the ground surveys. One of his assistants, Fred Jolliffe, would later be asked to map the Yellowknife area.

This frantic outburst of activity put an enormous strain on the transportation system. Explorers, missionaries, and fur traders had always travelled by dog team in winter, or by boat in summer. Regardless of the season, it was an arduous process from Waterways, Alberta, north down the Athabasca, Slave, and Mackenzie rivers towards the Arctic Ocean. The only major obstacle along the 1500-mile route was a sixteen-mile portage around a series of rapids that stretched between Fitzgerald and Fort Smith. This trail was simply known as 'the portage,' and it caused the same sort of bottleneck as the Chilcoot trail during the Yukon gold rush.

Passage on the water was hazardous but straightforward. Boatmen moved passengers and freight using York boats, shovel-nosed scows, and wood-burning paddle wheelers. There were two sets of these, operating above and below the portage. On land it was a different story. Two men, the Ryan brothers, had exclusive rights to move freight on the road they had built and maintained, and were quick to take advantage of the business opportunity.

Stan Graber While hanging around the railway yards in Edmonton, Tom saw a couple of Linn tractors on a flat car. Examining these, he was surprised to discover that they were the very tractors he had used in northern Manitoba. Tom decided, then and there, that he would ask where they were being shipped in the hopes of getting a job. He found their destination by following the Muskeg Express, a once-weekly railway through the woods to Fort McMurray. There, the tractors were unloaded and driven to their new owners in Fitzgerald.

Tom Payne I heard from Mussens, Ltd., the makers of Linn tractors, that 'Mickey' Ryan ran a freighting operation at the portage. He had purchased those same machines for a thousand dollars each. When I contacted him, I said that I was big and strong and could work like hell, so he hired me.

After the MacAlpine fiasco, the Hudson's Bay Company had tried again to use the Linns in the barrenlands, and had failed. The engines

*Moving a barge over The Portage, circa 1930's. Photo by R.E. Folinsbee.*

were badly beaten up, so I overhauled them during the winter in the Waterous machine shop in Edmonton with Shorty Crumb, Ryan's mechanic. We practically rebuilt them: new drive sprockets, new tracks and rollers, one sprung clutch, one bum radiator, and other assorted parts. In all, it must have strained Mickey's pocketbook.

ALICE PAYNE Pat and Mickey Ryan were two brothers who had originally come from Indiana in 1914. They were first attracted to the new opportunities created by the tar sands, near Fort McMurray. There they established themselves with one team of horses and hauled mail, fur, freight and travellers from Athabasca Landing. Soon, they expanded their services to Fitzgerald and purchased the hauling rights held by the Hudson's Bay Company, which was only too glad to be rid of the business.

Five years later, the Ryans moved their headquarters to the portage, where they improved the road. It was a tough job. The nearest gravel was 300 miles south, at Lac La Biche, which forced

*Moving a barge over The Portage, circa 1930's. Photo by R.E. Folinsbee.*

them to construct a corduroy road of sand and clay, reinforced with logs. Maintenance was expensive. Rebuilding it after heavy rainfalls cost about $10,000 a year.

By 1928, they owned four trucks, two tractors, a bus, and ten teams of horses. At their headquarters at the Halfway House, the Ryans kept a garden and a herd of shorthorn cattle to feed their crew. They also produced grain and hay, and built bunkhouses, a cookhouse and two big barns, plus a repair shop and smithy.

This investment totalled about $100,000. To protect it, Pat and Mickey obtained an exclusive franchise from the Alberta Government for moving all the freight, but by 1934, public outcry about their monopoly resulted in the building of a parallel road. The Ryans kept the franchise in return for maintaining the new road, but it didn't last.

TOM PAYNE I worked with Ryan's crew on the portage for a year and a half. The big barges were hoisted out of the water, towed through town, then refloated again on the other side of the rapids. They had to be treated carefully in case they snagged on the telegraph wires strung on poles along Fitzgerald's main streets. When it rained, most of the work stopped. The horses couldn't move in the mire, and the tractor treads slipped on the clay. It was hard work, and the accident risk was very high.

Our crew consisted of Shorty Crumb and Paddy Poitras, Mickey's brother-in-law, who was part Indian. Other skinners, like Joe Lacombe and Ralph Cameron, trapped in the off-season. For Steve Parlee, it was a summer job to pay his way through medical school. He became one of the chief baby-snatchers (obstetricians) in Edmonton. Ned Chisholm was another Englishman. We all became good friends. Years later, Ned and Steve hosted the party where I met Olga, my future wife!

ALICE PAYNE Jack Taylor also lived at the portage. He had come to Canada from Essex in 1924 and worked on the Hudson's Bay Company boats. His career included being a wildlife warden, and then an office administrator for the federal government. Jack had attended Felsted, Tom's old school, and his father was none other than the 'Jolly Old Taylor' of the White Hart Hotel at Braintree.

TOM PAYNE One New Year's Eve dance at Fort Smith, trappers came for hundreds of miles by dog team. Jack Taylor howled at the moon outside the hotel just like a husky! He danced the Red River Jig with a Mercredi girl known as the Black Diamond. Jack dressed up as the devil, with a long tail and spear tip. She tore the appendage right off his shirt, and on being asked what she had in her hand, said, "Just a piece of tail!"

It was a real do, and by midnight there was no liquor left. Around his neck Taylor wore a baby bottle, complete with a nipple, and filled with brandy. He had it tucked inside his devil suit, and kept producing it while dancing, giving his partner a suck. Needless to say he was popular. According to Ralph Cameron, there was a two-gallon limit on the amount of liquor you could have, but Jack could always get some more from the government warehouse.

Just when the party started to sober up, someone yelled "aeroplane." Everyone quit the dance and ran down to the landing field. Flares were lit, and every Model T Ford and truck shone its lights at the end of the runway, built of rolled and frozen muskeg. It turned out to be dear old Matt Berry, flying an old Fokker plane with a cargo of liquor! The dance continued well into the New Year and half the next day.

Northern life had its own set of moral standards. The Justice of the Peace, usually the acting District Agent, a local man, performed many of the marriage ceremonies, and Jack was often best man. Some of them went like this: He starts off reading the marriage lecture, everyone of course being under the influence including himself, and he turns over too many pages at once. Then he's lost. So he looks at the bridegroom: "Are you willing to marry this woman?" "Yes." And he looks at the bride and asks her the same question. "Yes." And then he says, "Now get the hell out of here, you are married!" Of course they weren't.

GLADYS TAYLOR It was a blanket wedding, all a fake for the girl's benefit. Sometimes they even did it for a policeman, who would take the woman back to the barracks. In the morning she wouldn't want to leave thinking they had been married the night before. Poor girls, some of them didn't know any better.

Tom Payne  In his spare time, the Justice of the Peace doubled as a dentist. He had a diamond point chisel, and jumped the teeth out by hitting the end of the chisel with a caulking hammer held between his fingers. Then he fitted you for a pair of dentures. You chewed on a puck of clay, which he shipped to Chicago, and back came a pair of black teeth. Iron! Johnny Berens, the pilot on the riverboat *Distributor*, and a lot of other Indians around Fort Smith, all wore iron dentures until the day they died.

Local entertainment was pretty scarce, which meant that we had to improvise a lot. The local police were a handy diversion. I had honed my skills on local constables in my youth, and talent wasn't lacking in the general populace. These incidents improved our morale and the police's interests were well served. We sharpened them up no end.

While I was working for the Ryans at the portage, Percy Round was run in for stealing a table. Percy, an Englishman, earned most of his money as a bootlegger. The RCMP sentenced him to the woodpile: 200 cords of wood to be sawn into eighteen-inch lengths, and then he was free to go. Well, he sawed them all twenty-four inches long instead, and walked around Fort Smith with a smile on his face from ear to ear whenever he came into contact with a policeman.

Percy had to take to the bush for awhile. When freeze-up came, the police had to apply for new stoves. The old ones had eighteen-inch fireboxes, which warmed their three-roomed shacks. They would have had to saw off six inches of every stick, far worse than sawing it to length in the first place. The RCMP spent quite awhile without heat, and Percy decided to hibernate.

Ralph Cameron  When spring came, Tom and I were sitting on the steps of the Mackenzie Hotel one day. You could tell what was on his mind because he taught me how to use a mortar and pestle to crush rock. Conversations always got round to the chances of finding a mine. Finally, he told the Ryans he was going to prospect for gold at Great Bear Lake.

Tom Payne  I learned prospecting and geology by going up the west shore of Hudson Bay. Chemistry and geology had always interested me at Felsted School; they were my best subjects. I had no degree, but I worked and consulted with some of the best hard rock geologists in Canada, and read one hell of a lot. I got my education the hard way, taking prospecting courses and working at the mines.

RALPH CAMERON Mickey was sorry to lose a good man on the tractors, and complained bitterly to Pat, who, with Billy Wilson, the bookkeeper, had agreed to grubstake Tom's venture. Mickey always said what money he had, he earned with his fists, and he didn't want it to disappear too fast. He was an American with an Al Capone attitude. He didn't trust anybody and always had to have a bodyguard or two, even though he'd once been a boxer. Mickey was a real schemer, but for the longest time, didn't know about Tom's grubstaking. It's harder to describe Billy Wilson. He was a short, stocky man who had his fingers in all sorts of things.

Pat Ryan was altogether different. He was a swell guy. Everyone thought he was deaf. You had to put your arms around his neck and talk in his ear. I still remember how he treated one young fellow who hired on to drive a truck. Right at the halfway camp, there was a turn in the road. It was quite sandy. Everyone tried to get up the hill without changing gears. Well, the kid made the turn, and dumped the whole load. There were boxes all over creation, and the truck came right off its wheels. It took a couple of days to clean it all up. When the kid came to the office to see Pat, he said, "I guess this is it for me", but Pat looked at him and said, "No, I don't think you should go, and I don't think you'll ever do that again." Pat's generosity must have led him into helping Tom.

ALICE PAYNE The term 'grubstake' probably originated with the California and Yukon gold rushes. It refers to the provisions or outfit given to a prospector in exchange for profit sharing. Even with support, you had to have a day job, or live off the land. Dad's didn't amount to much more than a promise to help him out if he found something interesting. Pat Ryan and Billy Wilson sometimes sent him money or set up his accounts. They also agreed to record any claims he might stake, an important task, considering they cost five dollars each.

Ralph Cameron That sort of life was hard. I wasn't interested in looking after horses all winter, or hauling wood, so I went back on the trapline. My wife, Laura, used to tease me and say I only came home to do my laundry. She wanted to go out with me into the bush, but I thought it was too rough for a woman. I endured every sort of hardship, leaving home with few possessions, and sleeping out in the open. I would often waste precious energy scrambling for dry wood to light a fire. To keep my dogs alive, I had to fish in order to feed them. They would eat two or three every night. I used to trap anything with hair on it, but prospecting was definitely not for me.

Tom Payne I bought my first miner's license on June 27, 1933, and jumped aboard the *Liard River*, a fair-sized schooner equipped with two Cummins diesel engines bound for Great Bear Lake. I ran the motors, or rather held the chief engineer's hand because, unlike him, I had no marine papers and he hadn't the foggiest idea what a diesel was all about. It annoyed me to have to clean the injectors and fix

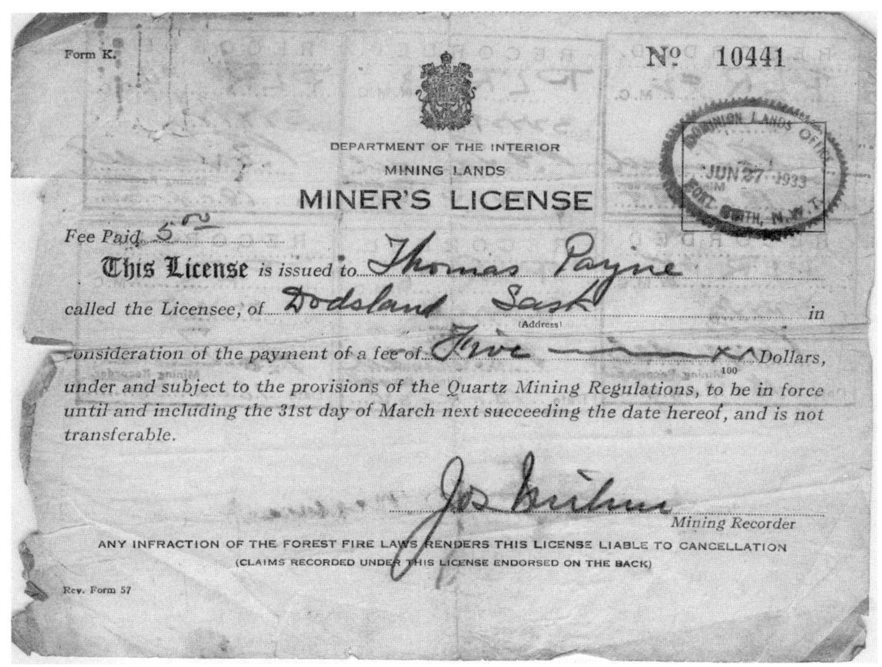

Form K. № 10441

DEPARTMENT OF THE INTERIOR
MINING LANDS
MINER'S LICENSE

Fee Paid 5.00

This License is issued to Thomas Payne

called the Licensee, of Dodsland Sask (Address) in consideration of the payment of a fee of Five — xx/100 Dollars, under and subject to the provisions of the Quartz Mining Regulations, to be in force until and including the 31st day of March next succeeding the date hereof, and is not transferable.

Jos Aitken
Mining Recorder

DOMINION LANDS OFFICE JUN 27 1933 FORT SMITH, N.W.T.

ANY INFRACTION OF THE FOREST FIRE LAWS RENDERS THIS LICENSE LIABLE TO CANCELLATION
(CLAIMS RECORDED UNDER THIS LICENSE ENDORSED ON THE BACK)

Rev. Form 57

*Tom Payne's first prospecting license, 1933.*

anything that went wrong and he got paid $500 per month while I only got seventy-five. The captain had never navigated rivers and lakes before, and if it hadn't been for the first mate, who ran the ship, we'd have gone nowhere.

En route, I had a chance to watch a memorable trial at Resolution. McKay Meikle was the presiding judge. The plaintiff was a widow. She always shacked up with the trapper or prospector who had the best winter grub pile for his trap line. A dance had been held in the biggest Indian whitewashed house, complete with home brew, fiddle music, and Red River Jigs. It was so cold outside that whenever you opened the door the steam came out! The widow said there had been a lot of partying, and she and her prospective partner had retired to the attic. While up there, she was raped, and was now in the family way.

The trial zigzagged back and forth, and the locals were most appreciative of the free entertainment. Finally Meikle asked why, if she was in all this terrible trouble, she didn't cry out for help. Her reply? "I was afraid that someone might hear me!" That did it. Poor old Meikle was flummoxed. His entire knowledge of court proceedings came from the Farmer's Friend Law Book, and the case was dismissed. After a good drink of milk and a box of soda crackers, he returned to Yellowknife.

ALICE PAYNE McKay Meikle was mining recorder for the Northwest Territories, first stationed at Fort Smith and later at Yellowknife. He was a likable and conscientious fellow. Interested in law, he returned from a trip to Ottawa as stipendiary magistrate for the Northwest Territories. His commission from Edward the VIII gave him "the rights, authority, privileges, profits and advantages" appropriate to his position, a very powerful mandate. Later, a legal background was required to do that job. According to Dad, the great thing about him was that if you lost your case and had lots of money, you could appeal to a higher court outside the territories and probably win.

TOM PAYNE By the time we reached Fort Norman, I had had enough of the *Liard River* so I jumped ship and joined the Hudson's Bay Company. I helped work the barges on the Bear River rapids, the fastest water I'd ever run. It was August and break-up was late. Icebergs were piled over the banks for fifty to sixty feet at the rapids, and a thin river ran down the middle. Now and then, a chunk of ice would slide into the drink, creating tidal waves that nearly swamped us.

One night the river started running pure ice! Great Bear Lake was emptying out with an east wind, which blew the ice downstream. Some pieces were as large as houses. All hands were needed to save the barge. We feverishly cut wooden logs and bored rope holes through them with an auger. These we bound together in mats and hung them over the sides. We stuck the spars, or larger trees, in the river bottom at a forty-degree angle to the bow. They were supposed to deflect the bigger chunks of ice as they whirled down with the current. But every twenty minutes or so, a ten-inch log would be severed just like a matchstick. Why the barge didn't sink, I'll never know. Our cargo was irreplaceable: seventy-five drums of oil and gas, a Case steam engine, and a sawmill with a Caterpillar tractor engine to run it.

*The Fort Resolution Pier. Schooners are the Fort Rae and the Speed II. The outermost sloop is probably the police boat Resolute, Aug. 1928. Courtesy of the John Russell/NWT Archives.*

Finally, Harry Weaver appeared with his barge. He was the best fastwater man ever born. We didn't have enough power to move upstream, so we took the huge drive wheels off the steam engine, wound a cable around the axle, then through running blocks in the bow, and fired up the boiler. This invention enabled us to winch our way up the 3500-foot stretch of limestone-benched rapids. With 150 pounds of steam, we pulled on that cable 'til it sang a tune, and oil dripped out of the center wick.

That cable was the only one around for a thousand miles. The Hudson's Bay Company lent it out on one condition: by taking it up as you went along, no one else could use it, so they could extract a toll out of anyone wanting to visit Bear Lake. They would ask the borrower, "What will it stand?" meaning how much cash have you got and how badly do you want it. Had it been left with a barrel anchored on the end, every power-equipped party could have availed themselves of it. No one up there missed the chance of making a dollar's profit.

We eventually succeeded in getting upriver, and anchored for a month while waiting for provisions to start up the sawmill on McVicar Arm. We were six workers and a cook. In the meantime, I ran a longboat up and down the rapids for a month. It was a tunnel boat called the *Dirty Dora*, with two opposed Kermath motors, and I made three go-round trips per day, with three and half tons per trip. It took half a day going up the rapids, and twenty minutes coming back!

Sometimes the boat would be standing still, and the water would go by so fast it would create a vacuum. The propeller wouldn't catch anything. You'd open both motors, pull them wide open, and let her go. When you struck bottom, you had to shut them off and let the waves pump you along. Then at the bottom of the rapids, you had to make a fast turn right in the middle and head for the cut bank. You had to be really quick or go over the boulders. It was very treacherous. You couldn't fool around or you'd be dead. There was no second chance.

When we got our grub and arrived at Great Bear Lake, we started out with the sawmill barge to Fort Franklin, just north of the mouth of the river. There I bought ten fifty-pound bags of flour from the Hudson's Bay Company with money I'd earned on the rapids. Five hundred pounds of flour at a dollar a pound. A lot of money in 1933. From the onset, I thought our winter grub pile looked too lean for seven men, and we thanked God for that flour before we got through.

We floated north along the shore to Fox Point, crossed over to McVicar Arm, and hitched up with the schooner, *Speed II*. More work. It was so soggy in the bottom you could drive a caulking tool through it just about anywhere. The motors were in bad shape and the valves were so corroded that when one stuck, you had to whap it to get it moving again. As the motor heated up, the valve covers had to be removed, and cold lake water used as coolant. After every trip, the crankcase had to be dismantled and the sludge scooped out by the handful. But we got our barge to the south end of McVicar Arm about freeze-up. Icebergs were still roaming the middle of the lake from the previous year. It was pretty bleak country.

Then the *Speed II* left for Fort Franklin to get the last barge of the season loaded with supplies for Murphy Services at Cameron Bay, outfitters at Great Bear Lake. Their scow was called the *Striped Assed Cat* and boasted ten-foot sides and a thirty-foot deck made of whip-sawn lumber, much bigger than the boat hauling it! It was loaded with gas and oil shipped by Imperial Oil on consignment to Murphy.

Vic Ingraham was a seasoned skipper. He was in partnership with Gerry Murphy and had a small crew which included an engineer and two deckhands. When a tremendous storm came up, the *Speed II* lost its compass bearings and was blown dangerously close to some rocks. A stir on Great Bear Lake is like a storm on the ocean. In the heavy sea, the huge barge might take a run and hit the boat, then rebound and snap. So they had to chop the towrope just in case.

When the pump-man accidentally dropped the lantern into the hot bilge, which had gasoline floating on top, the *Speed II* blew up! Exploded! Vic and Stu Currie, the other deckhand, tried in vain to rescue the two men sleeping in the bulkhead, but were forced to jump on a rubber raft before the ship sank. Vic was alive but badly burned, and the raft washed ashore. The untethered scow landed in the vicinity, and when it grounded, the seven men aboard scrambled to safety. They hunkered down in a makeshift bivouac for over a week, while two of them tramped a hundred miles southeast to Cameron Bay for help.

Harry Hayter, a bush pilot, flew his bright red Curtiss Robin all over Great Bear Lake during freeze-up looking for them. He sure as hell did a lot of tough flying, putting his own life on the line in that small plane. Harry was a damn good flyer. He had brought his plane out from New Brunswick two years before, and had rebuilt it himself many times over. Later he became General Manager of Mackenzie Air Service.

HARRY HAYTER It wasn't a fit time to fly, as the weather had turned nasty. But in those early days everyone took chances. They had to, to get things done. It was the in-between season and I'd just changed over from floats to skis. When the outfit didn't arrive in Cameron Bay, I decided to start an air search. My friend Ernie Mills joined me, and John Bythell of Canadian Airways flew the second plane.

I found out later that Vic and young Stu Currie had drifted like the others for eight to ten hours in minus thirty degrees F, and had landed on the Lake's northern shore. They believed their scow to be farther west when they'd cut it loose, so they walked, stumbled, and crawled the eighteen miles over land and ice to find their companions. It took forty-eight hours. In their dreadful condition it was nothing short of heroic.

When I flew over the scow, you could hardly see it because it was coated with ice. The only thing we spotted was a few tracks, which we traced to a little cabin in the only bunch of trees around there. We landed on a little muskeg lake, walked in, and found them. It was a pretty rough deal. We improvised a stretcher and carried Vic back to the plane. It was dark when we took off for home, but the whole settlement was there to meet us. Even Stan and Bill had finally arrived after their long walk. Earlier in the day, John Bythell had flown out the remaining survivors.

TOM PAYNE Vic was in bad shape, and his legs were frozen solid. He survived by living on drugs. There was a good surgeon at a frontier hospital in Aklavik, but gangrene had set in and Vic's legs were sawed off at the knees. He was flown to a clinic in Rochester, in the States, where several more inches of his legs were removed. He also lost most of his fingers, but as soon as he was able, he came home.

HARRY HAYTER Ingraham was a big man in every way: he had intelligence, imagination, and heart. Ernie and I would tell each other, "Vic will beat anything the Arctic can throw at him." He was a close friend. His wife and mine were also friends, and they had two of the cutest children you ever saw. In fact, most of the men on the boat and barge were friends. When a handful of people are thrown together in the barrens, they become comrades mighty fast.

TOM PAYNE Afterwards, Vic settled in Yellowknife and became the town's chief bootlegger. He operated from a small log shack, and had an assistant who was almost mute; all he could say was "Waw-waw" and hold up his fingers for customers to pay. A whisky bottle ordinarily cost ten dollars if you were sober, and twelve if you weren't.

All the liquor was flown in by plane, which landed at the Dogrib village on the east shore of the Bay. Stan McMillan and Harry Hayter flew in the crates. After unloading, they would take off and land at Yellowknife to be searched by the RCMP. The latter operation was a sham, as the police had to have a drink too! During the night, the liquor was dog-teamed to Vic's shack by Willie Wylie, Vic's brother-in-law, who was a trader from Fort Chipewyan. Waw-waw had it all sold by morning.

That first year, Vic made enough to save $10,000. He used his profits to build his first hotel and rooming house on the permafrost. It had ten-by-ten-foot mud sills, which had to be continually jacked up. Later Vic expanded, building a second hotel, and made a fortune. Dan Behan, an old wrestler, was his chucker-outer in the bars. Nothing was made to measure. Sometimes there was four inches of clearance under the bedroom doors; you could roll a beer bottle right into the next room. But the beds had clean crisp sheets and there was even a bellboy.

He ran such a good bootleg joint that when liquor became legal, everyone signed a petition for Vic to get a license. He made $3000 a night for the first few nights, and the head bargirl made enough to buy a trucking outfit in Edmonton. To get a room in Vic's hotel, you had to pay in advance and buy a bottle of whisky. If you didn't buy a bottle, you wouldn't get a room the next time. No damn chance, the hotel would be 'full up.'

Years later, I was sitting in the Yellowknife Hotel next to one of the local bible thumpers who fancied himself a medical expert. He didn't know me, and when Vic walked by, the reverend started to tell me all about him: "See that man there", he whispered, "You know what's the matter with him? He's got fallen arches." Hell, he didn't have fallen arches; Vic had two wooden feet! He always wore silk stockings over his beautifully carved legs, and sort of humped along on them. Everyone called him Old Cedarfoot.

Of course I wasn't aware of Vic's predicament at the time. Before the fire on the *Speed II*, we had ordered more food from Murphy Services. Harry Hayter was supposed to fly it in for us, but was too involved in the air search. We could hear him buzzing around as we unloaded the Case steam engine, the Cat, and supplies, and wondered why he didn't land. As daylight was getting shorter, we had to get ourselves organized.

There were lots of good spruce trees to build a log bunkhouse, cook shack, and log cache. We set up the mill and started sawing logs, which I pulled into the skidway on the go-devil, a sleigh made of hand-hewn birch logs. By December, our supplies were getting low.

On Christmas Eve we discovered that our grub cache was burning. The cook had left a lighted candle on a sack of sugar while getting some food out for the next day's breakfast. Everything went up in flames. There were ninety gallons of boat pitch in there too, and when the smoke cleared, all that remained were my ten sacks of charred flour and the fish net. That tainted flour and our ability to catch fresh water herring sustained us. To cap it all off, one of the men lost our fish net down an ice hole in the bay.

So we piled up some small poplar haystacks in hopes of snaring the odd rabbit that nibbled the shoots. I got ten rabbits ahead of the crew's daily rations, and then we had a council meeting. We decided to take the Cat, the tent, and the go-devil loaded with logs, and leave for Cameron Bay, some 300 miles away.

Our weak and worn out gang left on the morning of January 11, my forty-third birthday. We bridged thirty-five minor pressure ridges going up McVicar Arm, and retrieved some gas from the cache left by the ill-fated *Speed II* . As we headed out on the big lake, we traversed the major ridge that appears every year.

This barricade is formed when the ice piles up after it has shifted and broken up in the wind. It can extend the full length of the lake. I remember that the air was as clear as a bell, and when that ridge cracked, it sounded just like a naval gun. The ice beneath us also trembled just like an earthquake, due to expansion and contraction. One night in the tent, a pail of water standing on a powder box capsized and gave us quite a fright.

The idea was to travel beside the pressure ridge in order to find a narrow path to cross, even though the route was boring and monotonous. Meantime, I was on the verge of snow blindness. Not a bad dose, but I couldn't tell the difference between rough piles of ice and the hard snow banks. One side of my face was all frosty and felt stiff as a board.

We were moving along, when all of a sudden, *up jumped the devil!* I was so surprised that I thought the world had come to an end. But the apparition turned out to be Father Gathy, an Oblate priest! He was a big husky man with long whiskers, and a duffel bag tied to his wrists. Underneath his black Catholic robe, he wore two sets of Eskimo parkas and mukluks. One fur suit had the hair turned to the inside, the other had it outside. He had cocked up his komatik, laid his canvas over the sleigh, and spread out his goose-feather Chipewyan robe. Then he wrapped himself up, using his wheel dog for a pillow, and the rest of the dog team protecting his back.

A thick layer of snow had camouflaged his outfit until he was shocked and frightened into action. He must have heard our engine and felt the rumble of the Cat crawling along the ice towards him. I remember him still, standing there fifty feet ahead of the plow in bright moonlight. I don't know who got the biggest surprise!

ALICE PAYNE The Oblates of Mary Immaculate were pioneer missionary priests known to the Natives as Blackrobes. They travelled a huge territory stretching from the inhospitable shores of Great Slave Lake to the Mackenzie Delta, and made long journeys on snowshoes or in dog sleds, often sleeping under heaven's canopy in a deep snow dugout. They ate pemmican and frozen fish, and earned the respect of everyone by their struggle on behalf of the Christian faith. Father Gathy was particularly well loved. He was a gifted musician, and a master at card games.

Tom Payne Father Gathy and his dogs reached Cameron Bay before us and spread the news that we were out there. Soon a plane appeared from a hole in the clouds and buzzed us. Our next two days were spent crossing the worst damn pressure ridge I ever saw. The crumpled slabs were rammed against each other, piling chunks of ice in layers or standing them on end. Bits of open water and fresh ice made it all the more dangerous. We beat the sharp edges down with axes across our chosen spot, laid down all the trees and timber we had, and spent a day coating them with water to make a bridge. Next we decided to drive the Cat across, and see if we could pull the entire outfit over by cable.

I took the rope off my bedroll, tied it to the clutch lever, pulled the throttle wide open, pointed the Cat in the right direction, and yanked hard on the rope. She was on her own! I thought she was through the ice once, but she heaved and lurched sideways, and veered off across the lake! So I had to scramble to catch her. Thank goodness we had no more mishaps.

The following night we heard dogs howling. They sounded as if they were out on the ice about a mile from our campsite. In the north, where there are dogs, there are usually men and a chance of being rescued. We hurried to cut up a pile of logs, doused them with gas, and watched as huge flames shot skywards. This really got the dogs going. They made so much noise when they saw our fire that they woke up the police from Cameron Bay, who broke camp immediately and found us.

Thanks to Father Gathy, the police must have been on the lookout for us with an extra supply of food. Our ten rabbits were long gone, and we were down to the remnants of our dried fish, which had reached the throw-it-to-the-dogs stage. Now picture this: The police dogs were resting, and the lead sleigh driver pulled out his hunting knife, cut off thick wedges of bacon, and threw each dog a chunk. Bacon! Imagine wasting food like that with every starved man on that outfit watching.

Some of us were so weak we could hardly move. The police boiled up some strong coffee, which we sipped while watching them fry up large bacon rashers for us. We had to go easy on it though. We were so gaunt that to eat was almost too risky. It would have been wiser to have given us some soup and pilot biscuits first. Some of the men had excruciating pain from the stomach cramps.

From there to Cameron Bay, it became easier. However, I was in terrible physical shape from sitting on that iron horse for ten days and travelling over 300 miles, in minus forty to sixty degrees F weather. I had been forced to get up all night to baby the engine. We had the nose of the Cat stuck in the fly of our tent, and had to crank it every twenty minutes to keep it from stiffening. In the morning, the transmission oil, in spite of its dilution with coal oil, would be so viscous that I'd have to run the engine up and down the lake before it could pull the load.

As we left our campsite, one of our fellows came running up alongside the Cat and asked me, "What about the fire we left burning?" When I jokingly said, "Go on back and put it out," to my amazement he went back over a mile, extinguished it, then caught up with the outfit again. I felt pretty bad about that. Imagine, putting out your fire, even if it's on the ice 150 miles from shore! Such is the craziness of exhaustion.

After we got into town, we slept for twenty-four hours. Harry and Bunny Reed took good care of us, and one by one, we snapped out of it. It took a month for my face to get back to normal. My left cheek was a solid mass of scab and whiskers an inch thick. All the sticky juice had run out of my frozen, blistered face. What a mess. My whiskers pushed the scabs off, and I can still remember my new pink cheek.

Then the real problems happened; no one got paid. Murphy Services went bankrupt while we were out woodcutting. $75,000 worth of groceries was missing, which explained why we didn't get any supplies, and the Cat and sawmill outfit had lien notes against them. The government had even more liens against the lumber and when I asked for my pay, someone said, "What's that?" Mr. Justice Ives, in winding up the company, said it was "A very unfortunate state of affairs!" At least we had the Cat, lien notes or not, and I made a deal with Murphy to haul wood for the settlement for ten dollars a day plus board, payable every night in cash. Without that, I don't know what I would have done.

ALICE PAYNE Cameron Bay grew as a clearing house and supply center for the Great Bear Lake district. The wooded hills protected it from the fierce storms. By 1933, the settlement contained about twenty-five semi-permanent buildings and a shifting population of fifty to one hundred residents. There was a doctor, the RCMP, a Catholic mission, a couple of frontier hotels and general stores, a lumber mill, a government office, the mining recorder, and a post office. Harry Hayter

*Harry and Bunny Reed, in front of their Cameron Bay hotel. Photo by J.Taylor.*

and his wife had a house there, as did Harry and Bunny Reed, who operated one of the two wireless stations.

TOM PAYNE   While I lived at Cameron Bay, the Reeds owned a sixty by thirty-foot bunkhouse by night, and cookhouse and dining room by day. It was made of logs and had a sod roof. Harry grew a garden on top in the summer, and his pet eagle ate fish by the front door! There were two big oil drums, which were used as stoves, and a big square was hewed out of the floor around them. Inside, near the heat, it cost a dollar to unroll your Chipewyan robe to sleep. Near the outside walls it was only fifty cents, but your hair would freeze to the logs by morning. What a gang slept there: prospectors, priests, trappers, Indians, men, women. Everybody. It was the only accommodation there was.

HARRY HAYTER   My wife was only the third white woman to be that far north. Then Harry Reed brought his wife in, and later one other fellow, and then some people from Montreal, one with a wife and the other with his sister. Cameron Bay was a nice little settlement, and the police were a good bunch. We used to have a lot of fun with them. People would fly in to go prospecting or trapping. Some would stay two weeks, and I would either pick them up or fly in more supplies. I handled all of Great Bear Lake with my bush plane, and sometimes went as far north as the Arctic Ocean, or to Coppermine, or the Barrenlands.

The country was brand new then, and it was a rough, rough deal. At fifty or sixty degrees F below zero, there was no way we could keep the aeroplane engines warm. It wouldn't make any difference if we stopped for a week or just a few hours. We carried plumbers' blow torches, and converted them to shoot the flame straight up. It was a tricky business. When things started to get warm, there was fire danger if any gas was left in the crown. So we'd have to run the carburetors dry before turning the gas back on. It would take one to two hours to heat everything up. Then we'd have to act very quickly and get moving, because if it was really cold, we couldn't keep the engine running. We had to get the power, and get out, and GO! If we missed in starting, then we were in trouble. We'd have to start again from scratch. It had to be right the first time.

The cost of everything at Great Bear Lake was awfully high. When I started, freight was more than a dollar a pound to fly in. That meant a loaf of bread was at least a dollar. It did get a little cheaper once they had a restaurant. They flew stuff in bulk, and after break-up, more boats would arrive at the dock. The water freight rate was $300 per ton for 1350 miles. Caribou and fish supplemented the diet, and as usual, entertainment was provided by the police.

TOM PAYNE  A trial was held just before I came to Great Bear Lake. Joe Dillon, an Indian from Norman Wells, often in trouble, was accused of stealing gas from the RCMP. I think his sin was three barrels. Most of the time, Joe pitched his tent next to the barracks. Charlie St. Paul, Gilbert LaBine's partner, agreed to take the case for Dillon and told him not to say anything except "Don't know" or "Can't remember." Joe did as he was told. Two of the locals sat on the trial with Sergeant Baker and one other policeman. Baker was in charge of the Cameron Bay detachment. They could get nothing on Joe, who followed instructions and behaved like a mummy. They had to dismiss the case.

Baker was lying on a heap of fur robes piled in the corner of the Roman Catholic log church where the trial was held. He knew Joe had done it, and couldn't prove a thing. They gathered the papers on the card table, and the court and spectators, about twenty, which was almost the total population, started to move out. Baker said to Joe, "Who helped you load those three barrels of gas into your Chipewyan skiff?" Joe thought that the trial was over and he was free, so he replied to Baker, "Two other Indians." That did it! He should have kept quiet until he was right out the door. Joe was hauled back and put to work for the winter, cutting wood for Baker, and pulling fishnets.

At Cameron Bay, I met a wonderful Dogrib Indian and we travelled together for a year or two. One day we decided to go prospecting with another Indian guide to Hottah Lake, to the southeast of Great Bear Lake. We were looking for pitchblende, rumoured to be on a quartzite ridge east of there, and the guide apparently knew the way. We never did find the place.

Coming back, we hit the east shore of Hardisty Lake where we'd stashed our canoe. Here we had a big rumpus with the guide and parted ways. When we got back to Hottah Lake, there was that guy standing in the fly of his tent smoking a cigarette. He had come a hundred miles cross-country on foot, and had got there before us. That's eighty miles to get around the water, while we travelled night and day by canoe.

Later on there was a big pitchblende strike south of Hottah Lake, and that's where we should have looked. We figured that our Indian guide just wanted a holiday, with free food and lots of cigarettes. He ate all our grub and left us with a canoe, fish line, and a .22 rifle. My Dogrib friend perched on a rocky point, dangled a large hook with a ptarmigan leg lashed to it, and caught a fifty pound lake trout. That saved us.

Eventually we made it back to Boland's trading post on the North Arm of Great Bear Lake and worked on the *North Star* during freeze-up. A big snow and sleet storm caught us off guard. The open lake was rough, and we had to keep chopping the hatch open with axes. There was an awful load of ice on the two-inch ropes. How the top-heavy ship stayed right side up, I'll never know! Thank God the old Kermath motor kept going. My Dogrib partner wasn't at all worried and kept saying, "All right, I know," and "Star Harbour in half-hour," and sure enough in minutes we arrived, in calm water. What a relief! Like my old friend Stockings, this man never got lost.

After the *North Star* fracas we returned to Boland's with our boat and freight, borrowed a dog team for the winter, and returned to Hottah Lake. That was a rough trip too. My partner froze one of his lungs, and developed pneumonia. He nearly died from the infection. He was delirious for days, and I doctored him with steam and home-brew. But he lost his voice, and was afterwards known as Squeaky. When he finally coughed his way back to life, I said, "Where have you been to?" He said "To Heaven," and went on to describe what a lovely place it was. When I asked him how come he didn't stay, he replied, "They weren't ready for me!"

Squeaky was a good Roman Catholic, and he was worried about what would happen to him if he died with a bad conscience. One day he was awfully quiet, and I asked him what was the trouble. He didn't want to tell me at first, saying he had tried hard to keep the faith, but he'd done something he thought the church wouldn't like. He'd had a chance to say something to Father Gathy but was afraid. Now he was obsessed. So I said, "Look here, there isn't anybody here but you, me, and God. You confess right now, whatever it is. I won't say anything, and God won't either, and you can get it off your chest." Then he told me this story.

"It was a hard winter one year and I had run out of food. I caught a few fish and rabbits, finally I started on the dogs. In the end I gave up, and decided to find the RCMP. Out in the middle of a lake I found a crashed plane. It was all drifted in and looked as if it had been there awhile. The pilot was frozen to his seat. I took out my knife and pried bits off the plane for a great fire. Then I found a small stove, melted some ice, and drank the hot water. It was the first time I'd been warm for days, so I decided to make a shelter out of the plane and rest awhile. But I was desperately hungry, and seeing that pilot sitting there, all frozen up, I decided to cut a little piece off his backside. I thought then that it wouldn't matter. He saved my life when I was hungry. But now I'm afraid. I didn't tell the police, or Father Gathy."

So I said that the Blackrobes always say that your body doesn't matter to God, as long as your thoughts and deeds are honest. That pilot was already dead, and what happened to him doesn't matter now. When the ice melted in the spring, everything sank without a trace. You didn't kill him, and you have nothing to fear. You've got to eat, and you can't be blamed for that! Then I made the sign of the cross and said a prayer for him, just to make it official, and that was that. We never discussed it again, and he was content.

It's funny the way religion influenced these people. They would learn the Catechism by heart and as long as everything was okay, they could believe in it. I've seen both an Indian and an Eskimo who would attend the church school for years, learn English, and have all the veneer of civilization. Wear white man's clothes, and live in their culture. But the minute things got tough, they'd shed their new faith just like an extra skin, and go back to their traditional ways.

I also remember one man whose wife was very ill. There was no doctor for hundreds of miles and she was getting worse. They tried praying but it didn't help, so he called the medicine man who sang, chanted, and danced, and smoked up the place. This continued as long as the husband kept throwing beaver pelts onto the pile. The woman actually sat up and seemed to improve. But when the furs ran out, the medicine man packed up his kit and left. Then the woman fell back and died. I've seen a medicine man keep someone alive like this for twenty-four hours; not just alive, but getting better, just so long as the payment lasted. When he left, it was all over.

Personally, until I moved north, I never had much use for the church. When I was a boy at Felsted we had morning prayers before classes, and another dose in the evening. On Sunday, we prayed three times a day. That was enough for me. If I never went to church again, my batting average would still be good.

I felt the same way about preachers who say one thing and do another. On the farm in Saskatchewan I caught the flu and was bedridden for a week. I couldn't eat anything, and drank only water. No one came to see me, and finally I began to gain strength. I was barely back on my feet when there was a knock on the door. There was the local preacher saying, "Thought I'd drop in and see how you are, Tom, heard you had the flu?" So I said, *had* it is right, and now that I'm getting better, here you are. Now I can look after myself, and stand long enough to make a pot of tea, and now you show up. You just want to make sure I'm still alive so I can make a contribution to your paycheck. Well forget it! I got so mad that I nearly flaked out, and had a relapse!

But up north, it was different. I respected the Roman Catholic priests. They had a tough life, and often went to bed hungry themselves. They never refused to help you if they could.

# CHAPTER 5

ALICE PAYNE  As the Bear Lake prospectors began to fan out, Hottah Lake became the site of the next big staking rush. During the fall of 1933, the Indians found pitchblende in giant white quartz veins, east of the narrows of Beaverlodge Lake. This area had been partly staked by Joe Dillon in November, and at Christmastime, some iron-stained samples were shown to D'Arcy Arden, a local trapper, trader, and prospector. The vein, which could be traced for at least ten miles along a fault zone, varied from 175 to 250 feet wide. The Indians said that the white rock infilling reminded them of bone marrow.

FRANK CAMSELL  In Dogrib they call it Kweh Kha, which means rock fat, although the Bear Lake Indians probably called it Quin Kola. The dialects are slightly different.

ALICE PAYNE  In January, D'Arcy hitched up his dog team, headed south, and staked more than forty claims on top of the unpacked snow. When he returned to Cameron Bay, news of the find leaked out. One of the first to hear it was E.H. Hargreaves Jr., son of E.H. Hargreaves of Wright-Hargreaves. The senior man was then managing director of the Great Bear Development Company, which owned some silver claims near White Eagle Mines in the Camsell River area.

Young Hargreaves chartered a plane and flew in four assistants and a surveyor to set up a camp at Beaverlodge Lake. Compared to D'Arcy's two hundred mile trip on the ground, the flight was a real blessing. The crew found some reddish iron stain and staked on the snow to the east of Arden's posts. However, it wasn't until March that Hargreaves' men found a lens of solid pitchblende. They hurried back to Cameron Bay, negotiated with D'Arcy to secure an option on his claims, and took the first plane south to Edmonton.

One of the samples was an eight-inch square of solid black pitchblende streaked with canary yellow. The colour was the oxidation product of the uranium minerals, a valuable discovery. Major promotion followed and Hargreaves formed Hottah Lake Mines Limited to develop the find.

FRED JOLLIFFE As a government geologist, I was invited to look at the prospect. The vein was staked for several miles in each direction. A Gardner-Denver, two-drill, gasoline compressor was shipped to Rae by

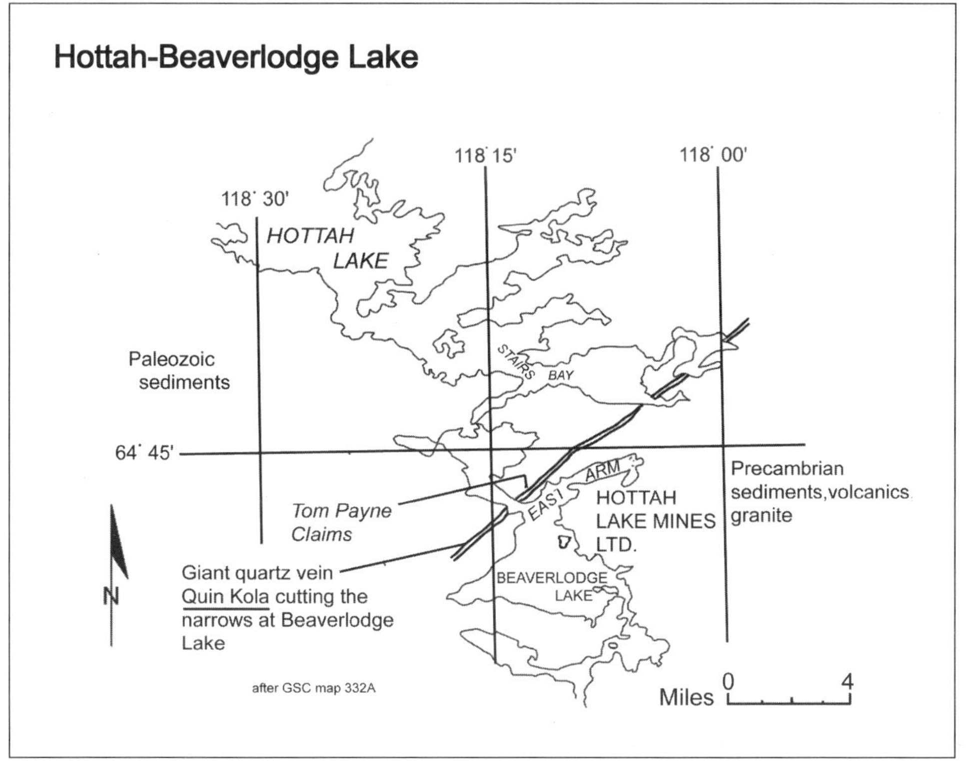

boat and then to Hottah Lake by air. By early summer, it was in operation. Two camp buildings had been constructed from local whip-sawn timber, and a crew of eight men set to work digging pits. Thirty bags of pitchblende were sacked and shipped to Ottawa for testing. My party chief, Dr. Kidd, later gave a talk about Great Bear Lake and compared the mineral deposits to a plum pudding: the ores were the plums, and the other rocks, the pudding.

ALICE PAYNE The first pit was sunk on the original discovery site (the plum) where pitchblende, with hematite, formed three lenses a few feet long. The hand-cobbed samples turned out to be high-grade ore, about one-third uranium. By the end of July, more headlines appeared in the *Edmonton Journal*. The prospect seemed guaranteed, backed by an established and well-funded company, and endowed with rich deposits. Along with several other optimistic prospectors, Dad packed his northern gear in April 1934. No one at the time could have predicted the abrupt end of the operation only a few months later.

TED CINNAMON Wilson and the Ryans told Tom that if he saw something interesting, they'd back him on a three way partnership. Tom wired them for credit, intending to stake claims at Beaverlodge Lake. The answer came back, "Give Tom Payne anything he orders and send us the bill." No one would believe it, so they had it confirmed. Tom knew nothing about how to stake claims back then, but his partner, Pete Lauder, was very experienced.

TOM PAYNE Pete Lauder was an American from the Ozark Mountains who happened to be living in Cameron Bay. He was an extraordinarily husky man, with a fifty-five inch chest and a bad hip. He called himself a frontiersman, but couldn't stand the sound of a train whistle. People said he was a fugitive from US justice. He'd been working on a river drive for Weyerhauser, a big logging firm out of St. Louis, and got into an argument with a co-worker. This man drove a pike pole through Pete's hip, but Pete grabbed it, yanked it out, and beat the other guy over the head "til he drowned!" I heard that later, at the age of eighty-five, he was still cutting railway ties with a broad axe in the BC forests. What a man! And a damn good prospector. Few could pan gold like he could.

We decided to take our meager supplies and go to Hottah Lake to stake on the snow alongside the claims group on the quartzite ridge. I had $600, which paid for an outfit, grub, and the plane ride, which was a Junkers owned by Con Farrel. The payload of furs from Coppermine lined the floor one bale thick. On top were seated a pimp, a prostitute, a bootlegger, a priest, and one dog team. Pete, Lou Parmenter the mechanic, and I completed the cargo. This was a typical northern group.

We had a hell of a whooping tail wind to Beaverlodge Lake. Hottah Lake was wash-boarded and we couldn't land, so we had to try the shoreline near the quartzite ridge. It was burned over and there were a lot of blackened two-inch poles sticking out of the ground. Just as we flattened out, we were caught by a fold in the wind capping the ridge. Poor old Con. He gunned the motor and gave all the rudder all he had. With a bump and kerplunk, we were down!

Lou Parmenter got to the door and yelled, "Get the hell out of here," and Pete did a nosedive into a deep snowdrift. All you could see were his feet sticking out. When he got up and looked around, he remarked, "God these planes land rough!" That had been his very first ride. We levered the plane with poles and hand spikes onto the ice, and straightened out the ailerons with single jack hammers. Finally, with some makeshift patches, and hay wire braces on the skis, Con Farrel took off for Edmonton.

Some fracas! It was lucky we were not all killed! The plane's propeller had come to a stop six feet away from a granite float boulder. I remember glancing at the pilot. Calm and cool, his face showed no expression. When we landed, the noise of the criss-crossed burnt poles scraping our tin fuselage was terrific. Thank God there was no fire. I was never so glad to have a layer of fur bales under my belly.

Pete and I staked onto D'Arcy's land, but didn't have enough money to record our thirty-six claims. We eventually got enough from the Ryan brothers, but with Pete around, we needn't have worried.

One day he was marking a claim stake when a man named Red McPhee appeared. Pete, hearing the snow crunch, turned around and Red started informing him that his work was no good. He had written his renewal license number on them instead of his original license number, as he should have done. Pete looked at Red and drove his axe into the trunk of a spruce tree above his head. He told him to beat it before he chopped him down. Red vanished and was never seen on Hottah Lake again.

Another day, Pete went down to the water's edge to get a pail of water. When he bent over, he looked just like a bear and I'm damned if someone from the other side of the lake didn't fire three 30-30 shots at him from out on the sand bar. So Pete grabbed his rifle and fired back. We never found out who it was.

Pete and I had a good camp. We made some extra money fishing for Bill Jewitt, who was flying to Bear Lake where Consolidated Mining and Smelting Limited were working on their claims northeast of LaBine's Eldorado property at Echo Bay. Every once in awhile, he landed and traded his grub for some of our fish: three and a half pounds for ten cents a piece. Bill had a good deal. Beautiful humpbacked whitefish are pretty good eating no matter where you are.

How he flew all that winter is a mystery. He was often above the clouds at Hottah, and we could hear him, but couldn't see him. Terrible weather grounded all the other planes, but it seemed that Bill flew every day hauling equipment for the mine. Pete and I reckoned that in order for him to take such a risk, he must have been flying out high-grade silver, not machinery.

BILL JEWITT Tom was right about that. I didn't know about pitchblende until I met Gilbert LaBine, then we, the Consolidated Mining and Smelting Company of Canada, later shortened to Cominco, went there too. But the depression really affected the company, especially in the vicinity of Great Bear Lake. Finally, in 1932, all outside speculations and joint ventures were dropped. But I persuaded Bill Archibald, who was in charge of exploration, that we might be able to ship out enough high-grade silver to pay our expenses. So in the spring of 1932, we flew out about nine tons of it.

By that time Eldorado was going strong, but that company didn't use planes for their pitchblende. They hired Indians to carry it out on their backs and load it on boats. Years later, many of those same men died from radiation poisoning.

TOM PAYNE Soon after we'd finished staking our claims, Pete was hired by J.D. Nicholson at the White Eagle Silver Mine. After one hell of a row, they split their outfit fifty-fifty. They even sawed their twenty-foot freighter canoe in half across the middle. Well, Pete knew enough to make a false end for his bow, and paddled his way out to Goldfields, on Lake Athabasca, which was basically a fish camp and the site of some claims he'd staked for the previous year. Eventually J.D. arrived in Goldfields too, where they made up again!

Pete prospected the rest of the summer and found a chunk of pitchblende weighing more than 150 pounds. He loaded it onto a plane and landed at my camp at Hottah Lake, C.O.D., along with two gunnysacks of old dirty underwear and some socks he'd stolen. I gave him a good going over for costing me $300. I had no money, and it took my entire bale of marten skins to pay off the pilot. Pete wanted me to help re-stake some expired claims, but I refused. We hadn't found

*The crew at Beaverlodge Lake. Tom Payne far right, wrapped in a blanket. Photo by L.B. Skinner.*

much at Hottah Lake, and back then, the ore was only good for medical purposes. This was the first smell of pitchblende that ever came out of Goldfields for many years, but it didn't do us any good.

In the spring, I did the assessment work on the Hottah Lake claims. Oscar Burnsted joined me. He was prospecting for Gilbert LaBine, but we were both short of supplies and decided to pool our resources and live together. In spite of a few caribou, the grub pile dwindled. Oscar's plane was supposed to come at break-up and retrieve him, but nobody came. He had a lot of sowbelly for bacon and rancid butter packed in salt. I had good butter in cans.

Oscar was a heavy butter consumer! He said to me one morning, after he'd taken half a pound out of the can, "Tom, go easy on it!" We had a bad row, but it blew over! Many years later, I saw him walking down Jasper Avenue in Edmonton. I yelled out of the car window, "Oscar! Go easy on the butter," and it stopped him in his tracks.

At break-up, we had a tame mallard that nested in the weeds on the shore of Hottah Lake. When we noticed that she'd laid three eggs, we had a great idea. We spotted them with an indelible pencil and every day she laid one more egg, which we smuggled for our hot cakes. Our trick produced more than thirty eggs, outlasting our meager supply of flour.

When the ice receded, we decided to build a raft and go up the lake to Ed Hargreaves' camp. On the way, we planned to stop at one of my caches. But we made an awful mistake by using dry logs. We had spliced them together, but they waterlogged easily and our raft slowly sunk in the ice water. We had to pile our worldly goods in the center to keep them dry, and our legs got wet above our knees. We poled along the shore all day and night, out of grub and starving.

To our dismay, on reaching the cache we discovered that someone had smoked all my tobacco and eaten all the emergency rations. Good thing that the planes had started to land, and Hargreaves, having plenty of food, was good enough to feed us. Oscar and I erected our tent and built a fire. We were chilled to the bone. He took water cramps, and rolled and groaned all over the floor. I gave him hell for putting up such a fuss, and in half an hour, I got 'em too! Cramps right up to your armpits, so I did my own rolling and cussing.

Eventually, Oscar left by McKenzie Air Service and I stayed to go prospecting. Claude Watt and Bert Olmsted were there too. They'd staked to the northeast of Hargreaves. Watt was living with a Métis, named Mary, when he made a find of pitch and went wild. She was also Mickey Ryan's sister-in-law. Watt wired her the famous note to Edmonton: "Start spending it, we've got it made! And she did! He landed up bankrupt.

In the meantime, I was hungry again. I knew Hargreaves' compressor operator gave him trouble, and one day he came to my camp and wanted me to fix it. "Here I am, starving slowly," I told him, "and you have the nerve to come down and ask me to fix the compressor so George can run it!" I told him that I'd take over, along with the hoisting, or the blasted thing could sit there idle 'til freeze-up for all I cared. I got the job. What a blessing.

There was really nothing the matter with the compressor outside of a valve stuck on open. Gas and oil were all it needed. It was a lovely piece of machinery. In my spare time, I hoisted at the shaft and helped Walter Rasmussen sharpen steel. But that was the worst job ever, as I was on the striking end of the sledgehammer.

LYMAN SKINNER Tom and I worked together when I was hired on as Hargreaves' bookkeeper. There were no books, just the monthly paychecks, and I spent the rest of my time helping the cook. My friend, Jimmy McBride, who was an RCMP officer in Cameron Bay, had promised me a job in the mines if I could get myself to Bear Lake. In the middle of the depression this looked pretty good, so I headed north. When I got to Rae, he sent word for me to fly into Beaverlodge Lake!

The camp had been operating full swing. Everything looked good. The frame buildings were up, there was plenty of food and equipment, and the shaft sinking was in progress. But no more ore was taken out of the ground, and soon the fate of the operation became clear. There was no money left and they folded. Not soon enough for some of the men. They were always the ones stuck with empty hands when deals went sour.

A month or two earlier, Hargreaves Sr. had arrived. As a special deal for the local people, he offered to sell some of the company's treasury stock, supposedly valued at forty cents per share, for twenty-five cents each. In Edmonton the same stuff was selling for eighteen cents, and a week later, it was worthless! No one at the mine suspected anything. They were told right up to the last day that things were wonderful, that it was only a matter of time until the big vein appeared. After all, the company had millions, and the operation was a very big deal. I didn't buy any stock, but some men spent two or three months' worth of pay and lost it all in a few weeks.

Just before the camp folded, Tom and a Swede had brought down ten tons of grub, their winter's supply, on a barge from White Eagle. Hargreaves had appointed the Swede and me as caretakers, but Tom missed out. He had been there longer than me, and was very disappointed. So Hargreaves said okay, if Tom wanted to stay, there could be three caretakers. Our main responsibility was to see that the Indians didn't clean out all the stores. As things turned out, Tom and the Swede were just as much of a problem.

*Lyman Skinner feeding the dogs outside the cook shack at the Hottah Lake Mines camp. Photo by L.B. Skinner.*

When the camp shut down, Tom moved into his log cabin, about half a mile from our camp. He stayed there a couple of months, but when the weather got desperately cold, he moved back in with us. He would have been crazy not to! That winter, on New Year's Eve, it was seventy-two degrees F below zero in Cameron Bay. We had no thermometer at Beaverlodge Lake but it must have been equally cold there. Several nails pulled out of our frame building, going off like .303 rifle shots. Once the temperature dips to minus 60 degrees, the frost shrinks the nails to the point where the wood lets go. Two or three went out on us one night. Frame buildings don't last too long up there!

Tom looked pretty awful. Terrible in fact! Dirty, unkempt, all beard, like a real prospector! No money, no nothing. The Ryans had grubstaked him, but in those days that didn't mean much. You might get a new pair of pants, a few dollars, and a plane ticket to where the action was. Once there, you had to live off the land. Although the three of us spent the winter together, we were not very congenial.

Living the way Tom did could make you cynical and sour. I could see that he'd had a good education and that he was well read, but it

*Sinking the shaft, Hottah Lake Mines Limited, before bankruptcy, 1934. Photo by L.B. Skinner.*

seemed to me he was nearing the end of his tether. In the spring he took sick and went out in March, but returned in April. He had phlebitis, I think, and his stomach was bad. I thought, "Oh no! Three of us again in the same shack," and when he got off the plane, I got on. That's the last I saw of him until years later, when he showed up in Ottawa and took me out to dinner.

ALICE PAYNE    It's curious that Dad remembered Hargreaves Camp as a lovely place! The main reason was likely the grub pile. The food, and its disposal, was probably why Skinner had been appointed caretaker. It was also the reason why Skinner didn't get along too well with the Indians, who would stand around for endless days waiting for a handout. Dad couldn't see why the extra food should not be shared or, preferably, sold.

TOM PAYNE    We wintered well in the midst of plenty. At Christmas, we invited Claude Watt and Mary, and their family, for dinner. We had lots of real good home brew, very high-class stuff with dried prunes, raisins, apricots, and some sugar. We made ten gallons of it. I'd taken the coil out of the compressor to distil it, and every drop of alcohol was double distilled and filtered through charcoal. I was an educated home brewer, having served my apprenticeship with Hargreaves' assayer!

Thanks to the bush pilots, we had chickens, apples, oranges, and Christmas pudding. When Watt and his family arrived by dog team, we all sat down to eat our fabulous feast. In the middle of it all, one of Mary's kids started howling real hard and wouldn't eat. All that trouble to get these special treats and the child wouldn't stop crying! When I asked Mary what was the trouble, she said, "He wants a fish head." We had frozen fish outside for our dogs and when she heard that, her eyes lit up. She rushed right out and got one, and the kids started sucking on it! Imagine that! Poor little guys wouldn't even touch an orange! Raised on fish, they weren't used to 'white' food.

HARRY HAYTER    The Indians used to eat the darndest things. Someone at Great Bear Lake gave me a little pup, a wire-haired terrier, and I was bringing him out of the north. It being nightfall, I stopped at a trapper's cabin somewhere along the way. He invited me to sit down and have a bite. I didn't know what sort of meat it was, but I gave a little piece to the dog and it got sick. So I asked him, "What in hell are you trying to do to me here? Even the dog can't cope with this stuff." He answered, "There's nothing wrong with that, it's good muskrat. I shot it just a few days ago." Well, I can rough it but there's a limit. In a pinch, I could usually find something edible in the plane.

TOM PAYNE I'll never forget the Indian woman I met near my Hottah Lake cabin. One night, a dog woke me up with a mournful howl. Outside, a woman's thin emaciated frame was leaning against a spruce tree, next to her dog team. Her snowshoes were all iced over, so I cut the laces, and took off her moccasins, giving her a pair of mine. She went back out and retrieved a bundle of caribou robes, inside of which was a small, bone-thin baby with its tiny fist in its mouth. What a mess! As luck would have it, I had two cans of dried milk, and I fed that baby drip by drip with a piece of cloth, she having no breast milk. You could watch him recover by the hour. I fed the mother on caribou soup, and her own milk returned within two days. A good thing, because the canned stuff was getting low.

Several days later, I learned her story. Locals called her One-eyed Annie. Her husband had recently died down at the narrows at Isabella Lake, southeast of Hottah Lake. She apparently had trouble delivering her baby, but had survived, although she was so weak she couldn't haul in her frozen fish net. So she donned her snowshoes, and left with their four dogs on the fourteen-mile trek to the settlement at Hottah. The snow was almost two feet deep, and the dogs were weak. She left one in a snowdrift, and had to beat the other three to keep them going. The last three miles, she broke her own trail.

After a week or ten days of chewing on caribou bones, Annie and her baby had improved enough to travel to the Indian settlement at Big Island, some fifty-five miles farther north on Hottah Lake. At the same time, I made a trip back to her tent to see if there was anything salvageable. It was still winter, and everything was frozen as if in a time capsule. Her dead husband was parked upright against a spruce tree in front of the tent fly. I tapped him with my axe handle, and he was as hard as an icicle. He looked as if he had just died of tuberculosis.

For years, Annie never passed my shack empty-handed. If I was away, she'd leave a lovely pair of mukluks and mitts on my table. They were beautifully beaded and sewn with sinew. As a young woman, a sharp pole had poked her eye out as she rode in a sleigh over a portage trail. The dogs were chasing a caribou, and it was too fast to stop. She kept that eye socket covered with a rag and draped her black hair over it to keep people from gawking. Even so, she was a sharp-looking woman!

That winter, the three of us racked our brains thinking about how to get our money back, along with last year's pay. I figured out that we should trade off supplies to the Indians for furs, as it looked wiser for us to settle with the company than waiting for them to deal with us.

Skinner was in a better position than me. His father in Ottawa could find a lawyer to try and recover his pay before bankruptcy proceedings were filed against Hottah Lake Mines Limited. But the Swede and I, we were hooked and before we got enough traded, down came Sergeant Baker from Cameron Bay and spiked a seizure notice on our grub cache and the buildings: Leigh Brintnell of MacKenzie Air Service had a large unpaid flying bill, and Constable Langfeldt arrived with Harry Hayter and the sheriff's bailiff to take inventory. We were beat! Skinner had already left by plane, so I took what food Baker had set aside for us, and went to the east end of the lake to trap marten and mink. Then in the spring, I flew out to Rae.

Hottah Lake was bad business. It didn't do anybody any good, even Hargreaves Sr. He made a little money, but one day he stopped for a shave at old man Watt's barbershop in Fort McMurray. Watt, then over seventy, was Claude Watt's dad, and the conversation drifted to Hottah Lake. Watt said, "I have a son up there who has some good pitchblende claims, but he's hard up and a man by the name of Hargreaves, who has plenty of money, is trying to starve him out. If that old bastard ever comes into this shop, so help me God I'll cut his throat!"

Watt was vigorously stropping his straight edge razor at the time, and damn near cut the strop in half. Hargreaves was terrified, and although he was only half-shaved with soap clinging to his face, he jumped out of the chair and beat it down the street! He was half loaded anyway, but the story made the rounds of Fort McMurray in short order.

# Yellowknife

## Chapter 6

Tom Payne When I left Hottah Lake I had some furs, and just enough money to fly out as far as Rae. I was terribly sick and couldn't keep a thing in my stomach except powdered milk and low bush cranberries. Lou Parmenter had to push me up the ladder of Walter Gilbert's old, single engine plane.

After we landed, I started to walk up to Jim Darwish's trading post and barely made it. I crawled over the two logs bridging the creek, and Mrs. Darwish, God bless her soul, took me in. She fed me lots of milk, rice, and sausage meat wrapped in bay leaves. It was a Syrian dish. Despite all I'd been through, with her help, I recovered.

Soon, the priest from Rae arrived with his schooner en route to Yellowknife. While there, his pilot and Red Hamilton, who was running the engine, got awful drunk on home brew. I was still recovering, but took his place and ran the boat to Yellowknife Bay.

Red worked for the priest, and was the same fellow that had fed the priest liquor one Christmas in Fort Resolution. They had him in a coffin, and Red was pouring the rum into him! The priest was dead drunk! Out cold! Not a movement in him, and all the Indians took their caps off and walked past the casket. Everybody was signing the cross. Well, the third day, he rose from the dead. Walked right around the damn settlement! That upset the Indians terribly. They all came in offering furs and God knows what.

When we arrived in Yellowknife, there were only a couple of tents in the bay. I went over to Burwash where they were getting the timber ready for collaring the shaft, and Ole Hagen hired me as compressor and hoist man.

Alice Payne Burwash Yellowknife Mines Limited was a subsidiary of Bear Exploration and Radium Limited (B.E.A.R.) that operated the silver mine at White Eagle. It was directed by Major L.T. "Lockie" Burwash, who sent several men out to prospect around Quyta and Prosperous Lakes. They found a little gold north of the Yellowknife area, and a small lens of high-grade gold quartz about twenty feet long on the east shore of Yellowknife Bay. There they staked the Rich Group in September 1934 while Dad was still at Hottah Lake.

Jack Moar I was the pilot who dropped Burwash off at Yellowknife Bay. We were headed north to Coppermine, where everyone was searching for the Franklin graves. On the way, I stopped to pick up some extra gas in Yellowknife. Punch Dickins told me he had cached two ten-gallon drums behind Mosher Island, and I could have them if need be. Well, I didn't really need them, but you always like to have extra gas, especially when you've never been in the area before.

I was just pouring the fuel into the wing tanks when my engineer hollered down from the cliff top. "Hey, you know there's gold up here!" He threw down some samples and you could see the free gold in them, but when I asked how big the veins were, he said, "Oh, just little, six or eight inches." So I told him we wouldn't bother, we'd just go on to Coppermine. In hindsight, those must have been the same veins Tom discovered.

On Bear Lake we stayed with the LaBines, and when Walter Gilbert came back with Burwash, I flew him and his crew back out. Burwash requested that we stop again in Yellowknife, so I asked him if he wanted to see the northwest shore behind Mosher Island. He declined my offer, even though I told him I'd seen some gold showings there. He commented, "Oh we have a lot better than that on the eastern side." So we let him off, and those claims, called the Rich group, never amounted to very much.

Alice Payne Burwash was the first to sink a shaft, but B.A. Blakeney made the initial gold discovery in 1898. He sent two samples, panned near the mouth of the Yellowknife River, to the Geological Survey for assay. Blakeney was one of the Klondikers who spent the winter on Great Slave Lake. One sample contained neither gold nor silver, but the other contained 2.158 ounces of gold per ton, and 0.408 ounces of silver. Anything over one ounce per ton is considered high grade.

In 1914, Dr. Charles Camsell examined the Pine Point lead-zinc deposits while conducting an exploration program from Lake Athabasca to Great Slave Lake. At Fort Resolution, he met two prospectors who had been across the lake to the Yellowknife area, and recounted their discovery of a small high-grade gold showing. Interest peaked further with the 1932 publication of a Geological Survey of Canada Report. A year later, the U.S. government went off the gold standard. Their currency was devalued, and gold rose from $23 to $35 per ounce.

When Dad arrived in Yellowknife in 1935, there were bush camps all over the area. Men from Great Bear Lake, Goldfields, Northern Manitoba, and Ontario hurried to stake their own claims. While Cominco's men continued to prospect at Walsh Lake, Burwash's crew, Baker and Muir, prospected the west side of Yellowknife Bay. They staked the Giant group, which contained a number of small high-grade veins. Other men, like Murdoch Mosher, staked on speculation, half way between the Rich claims and the Giant discoveries, near the island that would bear his name. Jack Moar landed his plane along this rocky shoreline.

Fred Jolliffe The federal government also had men out in the bush that summer. As part of the government's effort to combat the depression, the Honourable Wesley Gordon, Minister of Mines, gave a million dollar grant to the Geological Survey of Canada in 1935 to investigate. At $2.50 per man per day, 2000 men could be sent out into the field for one hundred days. Within two years, all the money was spent.

Alice Payne  Jolliffe had been with D.F. Kidd at Bear Lake in 1931 and 1932, and with Major Burwash at Yellowknife Bay in 1934. The following year, Jolliffe was appointed to lead the 10,000 square mile mapping and prospecting trip covering both regions. He arrived in the harbour after an exhausting six-week train and boat trip. The captain who ran the *SS Liard River* had his deep-sea papers but was a poor navigator. The declination up there is about 35 degrees off true north, and he corrected his compass the wrong way. The boat missed Fort Resolution and went to Reliance instead. This was likely the same captain that Dad had sailed with before.

Jolliffe set out with a crew of fifteen assistants, only half of whom had ever taken an elementary geology course, and only two of whom had ever been out in the bush. They had one map that showed the Yellowknife River and the shoreline of Great Slave Lake, and their job was to survey the rest. Both topographic and geologic maps were made with the most primitive tools, including some of Sir William Logan's original equipment. They dragged Logan's 'fish' behind their canoes to determine distance, counted their paddle strokes and paces over the tough portages, and used eyeball and crude triangulation methods.

Peering out the window of the bush plane, Jolliffe and his senior assistant sketched rough topographical maps. The junior assistant was often too airsick to do anything but moan. On the ground, two-person crews were dropped off 150 miles in the bush and told to canoe back by hand-drawn routes. More than thirty years later, Jolliffe remarked, "Poor fellows, thank God it was one of the last of those efforts."

Fred Jolliffe  One of my assistants, a non-geologist, was the same Lyman Skinner who had over-wintered with Tom one year at Beaverlodge Lake. Skinner had also traversed the country with Neil Campbell, a geologist who later became chief geologist for Cominco. Another man, Norman Jennejohn, had been in my field party at Great Bear Lake in 1932. He couldn't stand the flies. We had a contest once to see who had the most purple spots in one square inch, and when we counted up the bites, he had fifty-two! When he got back to university,

Jennejohn switched from geology to dentistry, but he had experience and I needed men like that, so in 1935, I asked him to be my sub-party chief.

When we moved our base camp by air to Rae and Russell Lake, we noticed a plug of light pink granite at Stock Lake. We had run a couple of traverses on the ground, hit the granite, and decided it was all granite, but from the air we could see the dark green volcanics wedged in behind it.

In early September, I sent Jennejohn down to finish the mapping. The next day three independent prospectors arrived: Vic Stevens, Ed McLellan, and Don McLaren. They announced that they had one more week to spend and asked if I had any suggestions. In those days it was never our policy up north to keep things secretive. Prospectors could study the current Geological Survey maps at any time. Anybody could. I recommended that they look near the granite plugs at Francois Lake and at Stock Lake, where I had sent Jennejohn.

They returned after three days, very excited, saying, "Francois River was no good, but look what we got west of Yellowknife Bay!" They dug into their sample bags and produced some rocks containing great gobs of gold!

VIC STEVENS I had worked my way across the country from Lake Athabasca to Great Slave Lake. There I met Tom, who was employed by Major Burwash. The day we arrived we met Dr. Jolliffe, and he suggested that the west bank of Yellowknife Bay had good prospects. We camped on the shore along with a couple of young lads from the Survey, and they undertook to show us a large quartz vein. On the way I noticed a blue quartz vein, and in picking into it, found some free gold. So the next day we got busy staking. We decided to stay in over freeze-up in order to get all the information we could.

FRED JOLLIFFE They asked me to keep it under my hat for awhile, which was okay too. That was standard practice if you were privy to a discovery. It was getting pretty close to winter and we had to retrieve our various parties and fly them out to Fort Smith.

When I met up with Jennejohn, I said, "That was a nice find the boys made, wasn't it?" He replied "Yes, and did they tell you I made it? Just a technicality of course, but they asked to join up with us and walk the traverse in from the west side of the graveyard draw. It was the last hike of the summer, and we stopped and whacked at all the rocks. Midway between the stock and the main contact with the volcanics, a quartz vein, measuring a foot wide and traceable for 300 feet, was found in a fracture zone up to four feet wide. The free gold was found over a length of twelve feet. I knocked off a chunk and showed it to Vic Stevens, and he borrowed the government axe and started to stake!"

Well, that was new! I didn't know what the hell to do. It was the first time I had ever had a party under my direction. In fact, the Survey had made the first discovery and here I was, keeping it a big secret. By this time, a week had elapsed. I thought of wiring Ottawa, and then everyone would know. I knew it was the end of the season, and at Fort Smith, there was a great horde of prospectors all coming back for the winter. So I decided to tell them the entire story. As we

*Two of Fred Jolliffe's assistants for the million dollar survey in 1935. L. B. Skinner and Neil Campbell stopped for lunch . Photo by L. B. Skinner.*

gathered around a table, I produced the samples. And boy, in no time flat, the air was full of planes all heading back to Yellowknife!

Bill Jewitt I was in charge of northern exploration for Cominco from Great Slave Lake to the North Pole, and happened to meet Jolliffe in Fort Smith. He asked me what I thought about the announcement. So I told him he was doing the proper thing, telling me and everyone else together. It had to be a first come, first served basis, not a private show. I got in touch by radio with Mike Finland, and he flew in with a crew to Kam Lake and staked the fourteen claims in the Con group. George and Jack Russell and David McCrea ran into Vic and the McLaren brothers going south. They leapfrogged each other and kept right on going.

Mike Finland That September, I was doing assessment work in a tent camp at Walsh Lake when I got word to fly back to Fort Resolution and meet Jewitt, who was going to Pine Point to close up for the winter. He knew about the gold at Yellowknife and I was to go there and see if we could tie onto the staking that had already taken place. So I picked up the three fellows at Walsh Lake and moved them down to Kam Lake.

I was both the pilot and the engineer for my little Gypsy Moth. It could make 85 mph on floats with no head wind. With the open cockpit, you had to have a full flying suit on, and you could only take two people, a lunch box, a few tools, and a bedroll. I had to bring in the crew one at a time, with a few pounds of freight. There was no choice of plane, the company had them and we used them. It gave Cominco a big advantage because everyone else had to go over land by water or on foot. We could get there first and stake any ground still left open.

Tom Payne While the big rumpus was on in Fort Smith, I was still at Burwash. The main showing was on the crest of a hill, about sixty-five feet above the lake level, and 2000 feet from shore. In July, a pit had been dug and sampled, and the assays ran at thirteen ounces of gold per ton. Phenomenal! By early September a vertical six by twelve foot shaft was being sunk to a depth of thirty feet.

I got on well with the McLarens, who were prospecting for Ventures, and then I met Vic Stevens and Ed McLennan. Vic had prospected from Lake Athabasca to Great Slave Lake. He had been with Dominion Explorers in 1929, but was located at Tavani, on the west coast of Hudson Bay, the hopeful destination of our failed Linn Cat train journey.

Vic and the McLarens had a camp in the graveyard draw and they had already staked some eighty claims. Someone in Jolliffe's party had tipped them off and they came to Burwash's camp, borrowed a canoe, and paddled across the bay.

Vic Stevens We told Major Burwash about our find, as he had been most helpful. Due to the lateness of the season, we lacked certain goods, which he kindly supplied. We were anxious to get as much information as possible, so we were working seven days a week. Shortly after the ice formed we had a visit from Tom. On seeing our showing he was much impressed, and after returning to his camp, requested some time off to do some staking of his own. When Burwash refused, Tom went anyway.

Tom Payne All of Yellowknife Bay was staked nearly solid. Not much ground was left. Con had got in and staked some claims, while

*The head frame at the Burwash Yellowknife Mines shaft. Photo by A.W. Jolliffe, courtesy GSC.*

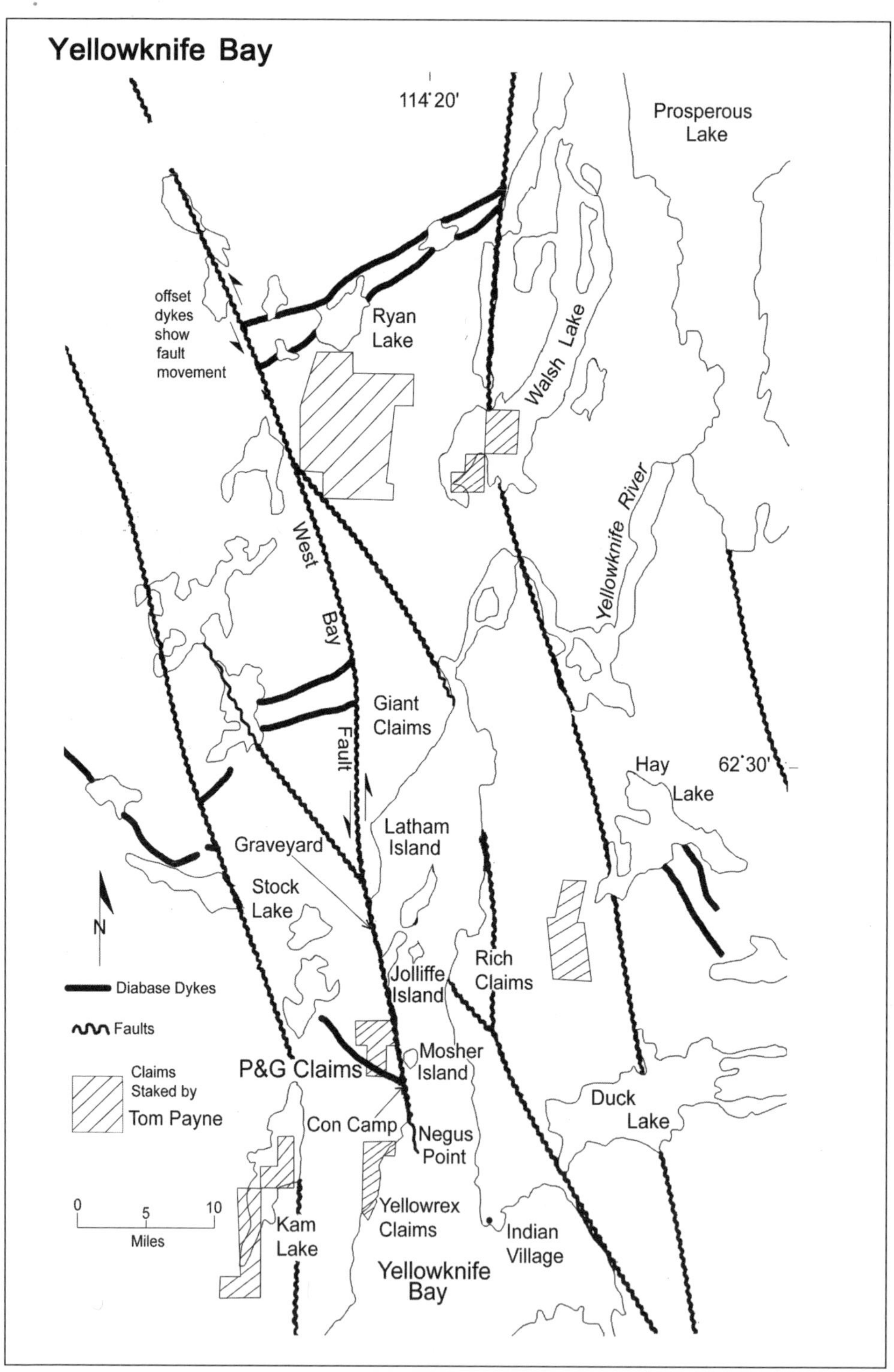
Yellowknife Bay
114°20'
Prosperous
Lake
offset
dykes
show
fault
movement
Ryan
Lake
Walsh Lake
Yellowknife River
West
Bay
Giant
Claims
Fault
Hay
Lake
62°30'
Latham
Island
Graveyard
Stock
Lake
N
Diabase Dykes
Faults
Jolliffe
Island
Rich
Claims
Claims
Staked by
Tom Payne
P&G Claims
Mosher
Island
Duck
Lake
Con Camp
Negus
Point
0
5
10
Miles
Kam
Lake
Yellowrex
Claims
Indian
Village
Yellowknife
Bay

B.E.A.R owned nearly everything else. Ole Hagen and I went down on a Sunday, and tossed a coin to see who would stake what. I won and elected to stake near the south end of Con's property, as Gordon McLaren told me they had a nice find on Kam Lake and to get as close as possible. Afterwards, these claims were known as the Yellowrex.

Ole Hagen staked north and got the future Negus Mine. Then we both got into a row with Burwash. Ole turned over his claims, but I wouldn't. I wanted it put in writing that I was free to prospect on Sundays, my day off. Burwash insisted that I was working for him the whole time, drawing their pay and eating their grub, and he was going to file a *lis pendant* (or pending litigation) against me.

BILL JEWITT I can imagine Tom would have a fight with Burwash; he was kind of a pompous guy and Tom was quite the opposite.

JOHN ANDERSON-THOMSON Tom was hauling plant machinery for Burwash up to the Giant property with a dog team. He was being paid with Giant stock, worth about two cents a share. That kept on for a while, but finally Tom said to the manager, "Look here, that's enough of this paper. I want to eat! Now this load is going to be paid in beans, flour and bacon or I'll pull it all back." He was pretty desperate and just about starving, so the guy broke down and produced the food.

TOM PAYNE Because I wouldn't turn over my claims like Ole Hagen, they refused to pay, feed or shelter me. I was barred from the cookhouse and bunkhouse so I took up residence in the yard, where a big spruce tree had blown down. I heated the root through with a log fire, then burrowed under the root, rammed my back up against the warm earth, and slept in my Chipewyan robe covered with a stolen canvas. It was bitterly cold, fifty below. Things got desperate. I ate off the slop pile for awhile, but they got wise and stopped throwing stuff on it. They even started calling the dogs into the lean-to for the leftover hot cakes from breakfast, so I couldn't get them. They just starved me out.

Then my big break came. Frank Camsell shot some caribou north of the Giant and left one buried in the snow for me. We had worked on the same shift together at Burwash's camp. I shook myself up, stole a pound of tea and coffee, some salt and flour, took my axe, 'inherited' some snowshoes, and headed for the mouth of the river and the portage to Walsh Lake.

The weather was miserable by the end of November. I retrieved the caribou, shoved as much of it as I could into my packsack, and staggered over to Walsh Lake where I staked sixteen more claims, the PRWs for Payne, Ryan, and Wilson. I hadn't the money to record them properly so I got the Ryans to do it, plus send me a box of grub. Years later, these claims, recorded the day before my forty-fifth birthday, in 1936, were sold for $262,000 cash and 300,000 shares in a three million dollar company.

When I had finished staking, I hit out for Burwash's camp and the old root shelter, and waited to get a plane out. Here I got into a real do. 'One Round McQueen' had jumped some of the claims I had staked for the priest in Rae. McQueen was a tough customer; no one could last one round with him in a boxing ring. We got into a row and McQueen dove at me, but I slipped him a lucky punch and had him groggy. I then proceeded to put him away until Jim Darwish's brother jumped on my back from behind yelling, "Don't kill the man." McQueen had had enough. He ran for the cook shack and anchored himself under the table before I could shake off the other man. Just then Burwash turned up. I told him he was an old man, but he'd better look out as I was not about to respect age in my current frame of mind.

Well, it ended up that I signed over the Yellowrex claims to B.E.A.R. for a ten percent non-assessable, non-divisible royalty interest. So I quit, collected my pay, and left on a plane for Fort Smith. I was so damn mad. I didn't know that in the long run, Cominco would end up owning both the Yellowrex and the Negus properties.

I made a deal with Joe Lanouette at the Smith Hotel. He was about in the seventh stage of alcoholism at the time. I agreed to cook and manage the hotel for him for the rest of the winter. I took two weeks leave of absence in March to guide Leigh Brintnell's men in staking the Akaitcho claims north of Giant. At the same time, I staked the Ann claims, next to the PRWs, and returned to Fort Smith until spring. Good grub and a warm cot put the life back into me.

Alice Payne  Dad didn't do too badly as a cook. He became an expert at making soup, and one day made a special pot for Paul Trudel, the mining recorder. Paul said it was the best he'd ever tasted, but didn't find out until later that it was muskrat.

While Dad was spending the rest of his warm, well-fed winter at Fort Smith, Cominco began to examine their property at Yellowknife. A number of narrow high-grade veins were discovered, and Mike Finland's crew began diamond drilling and trenching. They set up their camp near the lake at the foot of the draw. From there, prospecting and development work was still a lengthy walk every day over the rocks.

Tom Payne  I returned to Yellowknife in the late spring of 1936 and staked several more claims groups for the Ryans. One day I ran into Dave McCrae, who had staked the Con. He was working on one of the shear zones, making drill holes by hand-steel in order to blast out pits for sampling. "Come down on Sunday," said Dave, "I'll be up in the blacksmith shop pointing up some steel. There won't be anyone around, and I'll show you what we have there in the pits."

I took him up on the offer and slipped my canoe down to the island opposite the Con camp. That canoe was only twelve feet long and specially made to tie onto the fuselage of an airplane. But you couldn't spit out of the damn thing or it would tip over. It might as well have been made out of birch bark. No one would ever think of going anywhere in a boat like that nowadays.

I reached the north end of the island, later named after Mosher, and decided to wait because the winds blowing off Great Slave Lake made it too risky to cross. The water between the island and the mainland was like a mill race: fast and turbulent with many white caps. When the wind changed, and there was a lull between me and the school draw, I paddled over to a nick next to a fault and pulled my canoe up on the steep banks. The canoe was so light that you could hoist it with one finger, but I needed to weigh it down with rocks to keep it from blowing away.

On the way to the Con camp, behind what is now the old powder house, I came across a rusty gossan shear zone, which I picked away at with no result. Then I went on to see McCrae who was sharpening some steel getting ready for Monday morning. He showed me everything they had, and I promised myself that I'd go back to do more investigating because my little shear zone looked better than anything he'd shown to me.

So I strayed from the path a bit, and ran into it again. At first I couldn't find anything and was pretty disappointed. It appeared to be exactly like the Con show. But since the quartz looked so similar, I dug down in the rust until I came near to quartz. Filling my pockets with this crumbly gossan stuff, I scooped out another handful to see if it panned.

I had a large Yukon pan in the bow of my canoe, and I dumped the contents of my pockets into it, added some water and shook it up. I wasn't particularly careful with it. I must admit that I just stupidly surged it under the water pretty roughly. And my God, I cocked the pan up and the bottom rattled like shot. GOLD! I was stunned. Couldn't believe it. And just like everybody else, I had passed right by it the first time.

Everyday people stepped right over that vein on their way to the cook shack, but nobody paid any attention, even though the area had been staked before. Con was busy working on their ground and the path between there and the cook shack bisected my vein. It was a skinny section, about six to eight inches in a shear zone, but I couldn't take a chance. I went back and spent hours covering it up and left it just as God made it. I packed moss, windfalls, and little sticks into it, and covered up my boot marks with more twigs and leaves to make it invisible.

Now came the big problem: who in hell owned these claims? I wrote my friend, Paul Trudel at Fort Smith and he replied that Alex Mosher, a prospector from the east, had staked them the year before. No development work had been listed, so the current claims would expire and reopen for staking in three months.

What a long stretch. Everyone kept asking me why I was sticking around. All I could say was that I was waiting for the caribou to come, and act as crazy as possible. I was worried stiff. One flick of an eyebrow would do it. There must have been about thirty discoveries of free gold out in the bush around Yellowknife and at Ptarmigan, so everyone was out beating the bushes for more. When the news leaked out of a find by Don Cameron and the McLaren brothers, who were prospecting for the Mining Corporation and the A.X. Syndicate, everyone headed northeast to Gordon Lake. Everyone, that is, but me. I didn't have much trouble convincing people I was screwy. They thought that I was really 'bushed' this time.

I knew that I needed some extra help and wrote to Jack Taylor and Joe Lacombe, at Fort Smith, with no luck. Then to Jack Stevens at Bear Lake, but he'd gone to Goldfields along with my old partner Pete Lauder. No one seemed interested. My tent was opposite the hump, over by the graveyard, where Vic Stevens and Ed McLellan had camped. I made a cache and prepared myself to wait it out. I read and re-read the rulebook, planning the way to be first. The timing looked very tight to me, but I knew I had it right. Frank Camsell had tried to restake the claims, but his paperwork was faulty. You got a month's grace in those days, and he hadn't waited the thirty days.

VIC STEVENS After break-up we went prospecting on the islands at Great Slave Lake where we found gold in a large quartz vein. Because our assays were done in Yellowknife, we returned to our old camping spot and found Tom as the only occupant. The other prospectors had pulled out and gone to Gordon Lake to work on the Cameron show. While waiting for the assay results, I had long chats with Tom and his anxiety became very obvious. It wasn't too long before he blurted out, "Suppose you are after the same thing that I am." I had no idea what he meant, but I answered, "It could be." Whereupon he said, "I knew it. What should we do?" He then told me the full story and suggested we go fifty-fifty in the staking. He said there were others in Yellowknife who knew that the claims would come open and, like him, were just biding their time.

I reassured him, saying that I had known nothing of the situation. In any case, we weren't interested as we had 105 claims already and the company thought so little of them they had pulled out. Any more would not be welcomed. Tom was much relieved by this news, and asked me if I knew the best method of beating the others to the spot.

He had good reason to worry, as other prospectors had been hanging on all winter and were just as desperate. He felt that the moment they saw his canoe take off, they would follow and there would be a scramble.

I suggested he forget the canoe, as he could get there just as quick by land. The thing to do was high tail it down there, make and mark up his posts, and hide them near the corners. On the stroke of midnight, when the claims reopened, he could race down the lines, propping up the posts as he went. He followed my advice to the letter.

TOM PAYNE  I spent the next few weeks fashioning my posts by squaring off the faces and marking them. Then I hid them in the bushes and covered up my tracks as best I could. It was an awful job. God! To avoid being spotted you couldn't even smoke, and the flies were thicker than hell – black flies. It was just as bad as Bear River, where I'd seen people so bitten up you couldn't see their ears, only where they should have been.

I had a fishnet, and a few gulls' eggs, and a damned old tent! A bear tore half of it out! He just walked away with it, and I had to sleep under the tattered bit he left behind. The Ryans had grubstaked me, but no money had changed hands. And no one would lend me a dollar, except for Frank Camsell. We went out together for two weeks, and I told him that I'd split my money with him if I ever got any from the Ryans. And then when I did get a cheque, I told him he owned half of fourteen dollars!

One of the bush pilots, Con Farrel, offered me fifty bucks to stake those claims but I said, "Look, Farrel, you must be prepared to lose your money because we're gambling to beat hell." He took back the cash and I never felt so sick in my life. Hell, I could have owned Yellowknife for fifty dollars then! I damn near did own it in the end, but not quite!

ALICE PAYNE  While he was waiting, Dad checked his miner's license, first issued on June 27, 1933, at Fort Smith, for five dollars. It had been renewed in 1936 for one year starting in April. This put him in good standing. Dad could stake and record six claims for himself and six claims for each of two other licensees (proxies), or eighteen altogether. If the licenses expired, the owner forfeited his claims, unless a lease or patent was issued.

The number one post carried the name of the claim, the licensee, and mining license number. It also documented the date and hour of staking. Posts two, three, and four did not require as much information. Provision was made in the regulations for staking water claims by means of witness posts, describing the true location of the corner post. All of the details were sent to the mining recorder at the time of application. Small trees and brush were trimmed away from the boundary so the claim was clearly visible, and larger trees were blazed down the lines.

Once a claim was staked, and the recorder issued the identification tags for the posts, the licensee had one year to perform assessment work of one hundred dollars. The claims lapsed if this work was not recorded within one month of the expiry date. The regulation that bothered Dad the most was one that allowed the Department of Mines to grant extensions for periods of up to one year.

TOM PAYNE If Mosher had heard about my find, he could have applied for and got extra time to do his assessment, and I would have been out of luck.

JOE RANKIN Mosher had come to Yellowknife because of the Burwash find, and returned again in June of 1935 to prospect on his claims. He had another guy working for him and the two men prospected for a couple of weeks and left. Because they didn't find anything, no assessment work was ever recorded.

Tom went to the RCMP station before the claims came open. They used to get the official time from the radio signals people. Tom had the foresight to go to the police and check his watch. He was awfully goddamned poor in those days, but he had a good timepiece. And he said, right there in front of Al Fenton, "I'm setting my watch because I want to have the right time." In case anyone else turned up, he was going to ask who else had checked with the official timekeeper. If they put their posts up before Tom did, they would be illegal, and if they put them up afterwards, it would be too late, because he had already staked. He had a psychological head start.

ALICE PAYNE The night before Mosher's claims came open, Harry Hayter was in Yellowknife and Dad asked him to assist in the staking, but Harry was too busy flying. Then Dad went over to Pete Racine's Corona Inn, and persuaded Gordon Latham to help him for a ten percent interest. Gordon didn't know anything about prospecting but that didn't matter. The posts were marked and ready for action. He was still in his teens, but he was healthy, willing, and all that Dad needed him to be. Gordon had no vested interest in the claims, a qualification that eliminated every other able-bodied men left in Yellowknife.

TOM PAYNE As my watch ticked toward midnight we sneaked through the trees in silence, each to our own number-one post location. Then we sat and waited. On the dot of midnight, I jumped up and set up the first post. I fired my gun to signal Gordon to do the same. Then we both ran like hell down the 1500 feet to the next set and stood them up. We put up the other posts later. Our claims were grouped like the letter 'T.' Gordon staked the northern two and I staked the southern two. We would have had six claims but we missed putting up witness posts for the water claims to the east, out in the drink. We only got four claims, but they were the best ones. We called them the P&G claims group, for Payne and Gordon.

By that time, other men had arrived and we had done all we could. I decided to go down to the Con camp for breakfast, and as I sat in the cook shack with Cominco's men, a few of them were talking about the claims that were due to come open, right next door. They decided that it might be worthwhile to look into staking them. So I stood up and said, "Gentlemen, it's done!" They all gazed at me and laughed saying, "Tom, you crazy fool! Your staking is no good – they aren't open yet!"' They wouldn't believe it but when they checked later on, they found out it was the truth!

I never told a soul about the gold on these claims until three weeks had passed. Meanwhile I photographed the posts, cut the lines out good and strong, and got my A and B application forms back from Fort

Smith. When I got my forms and tags, I knew the claims were mine. The Ryans had put up the money to record them and I was safe. The date, August 26, 1936, was stamped on the back of my license.

With the paperwork done, I started poking around. I still had no money and precious little in the way of provisions: some jerky, flour and baking soda, my fish net, and seventy-five gull eggs. I was keeping them on ice, in a box, in a fissure crack at the back of the old graveyard across the shore from Weaver's trading post. The crack was overgrown with moss, and fifteen feet deep; cold all year round because of the permafrost.

Everyone in the old days used the crack to store their caribou meat and perishables. You attached your bundle to a piece of line, and then tied the other end to a stick with your name on it. And as for gold, even though someone had found a deposit in an ice lens at Giant, there was no fortune in my refrigerator.

Reliable help was expensive, but I knew an Indian boy who coveted my old .30-30 Winchester. So I asked him to help me blast out my ore pits for sampling, and then he'd own the rifle. I didn't even have enough food to feed him, so I'd have lunch with George Carter, the cook at the Con camp, and ask him for a couple of extra sausages. We made out okay.

Another man, Charlie Hershman, who was high grading at the Giant property, donated an old box of powder, some fuse clippings, and some two dozen 7/8 inch drill steels. I had enough money to buy some caps, and away we went. A word about the powder: the boxes had stood on end for so long without being turned that all the nitroglycerine lay on one side. It was risky stuff to handle; sixty percent and liable to explode any-time, but it worked fine.

The first pit we blasted was right on Con's work trail just below a little rise near their inclined shaft. Bob Armstrong damn near fell into it walking back ahead of the crew! We blasted out about half a dozen pits, then gathered up some old potato sacks and doubled them one inside the other. We filled them with the muck from the pits, and labelled each one according to its source. There were two tons in all, and the Indian boy and I got it all down to the water's edge a mile away, ready for shipping to Edmonton for assay.

I put half of it on Weaver's barge, and half of it on the Hudson's Bay Company barge, which was almost the last one out of Yellowknife Bay before freeze-up. Weaver's barge sank at Outpost Island, along with our sacks of gold ore, but the Hudson's Bay barge made it out. Its cargo created a furore when it was assayed at the University of Alberta. The main vein, which we traced for fifteen hundred feet, was worth $700.38 per ton or better than twenty ounces of gold per ton. It was high-grade stuff, and then I knew I had a real mine!

That winter we formed a small one hundred share company to consolidate our interests, and entered the information in the mining recorder's book as follows: G.G. Latham, ten shares; Tom Payne, thirty shares; G.P. Ryan, twenty shares; M.L. Ryan, twenty shares; W.R. Wilson, twenty shares.

We called our company Quin Kola Gold Mines Limited, after the Dogrib Indian name for quartz: Quin meaning fat or marrow, and Kola for rock. The word came from the narrows on Beaverlodge Lake near my old campsite, where a giant white vein cuts across the middle of the lake. Resources are the fat of the land, and my quartz vein certainly qualified. Finally I could look forward to a clean meal instead of scraps, and a real bed instead of sleeping in a tree.

# Ryan Gold Mines

## Chapter 7

Alice Payne It was important for Dad to do development work on his claims before there were any offers from larger companies. He was sure that the property would prove to be a mine, and he didn't want to sell too cheaply. This was his one big strike and it wasn't easy to hang on. The Ryans wouldn't put any money into Quin Kola, and Tom had to get the money from J.A. 'Jim' Pawsey in England. Jim's daughter, Rosemary, was once married to Bill Payne, Dad's brother, the physician and surgeon.

David Payne My mother, Rosemary, asked the family if they could help Tom to prove up his claims. Apparently when Jim gave the money to Tom he said, "This is a gift, and I don't expect you to repay it. But if your hopes come true, don't forget us." I don't know how much was involved, but we were not forgotten. I think my grandfather must have felt an affinity to things Canadian as four of his brothers had homesteaded at Edgerton, Alberta.

Nell Wheeler Tom later paid it all back, and double. You couldn't talk him out of it. He said that the loan meant the difference between hanging on or losing out, and there weren't too many people you could ask money from in those days. Tom was so appreciative. He was fair and honest. An exceptional man. When Jim was eighty years old, he and Rosemary traveled to Canada just so they could meet Tom in person.

BILL JEWITT A little group in Edmonton staked Tom so that he could work on the claims. It was the smallest grubstake anyone ever had. Bob Armstrong was up there, and helped out a bit. It was freeze-up, and already cold. Tom had nothing, not even a tent, just a tarp.

STEVE YANIK But the clothes he wore! He would try to patch them up with a needle and thread, his socks and everything else. Because that was all he had, and they simply wore out. And they weren't very clean either. There was no way of washing or anything else, you know. My wife remembers Tom coming in from the bush. His breeches were too tight, torn, and greasy, there was bare leg showing, and the seat was missing. Harry Hayter and a few of us were having a poker game, and we all put in a dollar to buy him another shirt so he wouldn't freeze to death. That was fairly common practice in the north.

BOB ARMSTRONG I arrived in Yellowknife in September 1936. The company suggested I pack only my shirt and a bedroll, but I ended up staying seven years. Bill Jewitt flew me in, ostensibly to sample the trenches, make a ground map, and get an assay plan going. Dave McCrae and his crew had prospected the Con group of claims, and found a number of veins. The highest grade was on the Con 10 claim, right next to the Con 4. Then George Kilburn, the head of exploration from Trail, came in to size things up. He decided right then to sink a little shaft, a decline on the Con 10 vein system. I stayed on to help.

Tom's holdings were pretty close by to the Con camp. The first time I saw him, he had a little wee boat, with a very small motor, an outboard, but all I could see coming across the lake was him sitting on the water! After he staked his claims, he used to come down for lunch at our old camp in the Negus draw. There were no hard feelings about Tom staking his claims. We all knew the tags on that property were going to lapse, but we were not as quick on the trigger.

I was in a tent down in the draw, and there was a little bridge across a creek. One day Tom came running down the trail full tilt, across the bridge, clomp, clomp, clomp, and banged on my door. He had a big chunk of rock with quite a lot of gold in it and said, "Christ! It's worth millions!" And he was right.

TOM PAYNE In the spring, the Ryans decided they wouldn't sink any more money into the ground. The few dollars they had invested in my mining licenses and in recording the claims were already too much. I could have bought them out for a hundred dollars, but I didn't have it. Pat Ryan and Billy Wilson had grubstaked me all those years, but Mickey Ryan had no faith and wouldn't help any more. The only thing we could do was form a new million-share company, take out 400,000 shares for ourselves, and sell another 100,000 shares to the public at fifty cents each.

ALICE PAYNE Quin Kola Gold Mines Limited held all the claims that the Ryans had recorded for Tom while they grubstaked him. That included the four P&G claims adjacent to the Con holdings, and three more claims groups on both sides of Yellowknife Bay.

All these properties were transferred into the new company, called Ryan Gold Mines Limited, which was incorporated on February 8th, 1937. Mickey Ryan was persuaded to act as President, Tom was Vice-President, and Pat Ryan was Treasurer. Billy Wilson became a Director, along with Moe Lieberman, an Edmonton lawyer who had tracked down and purchased Gordon Latham's shares.

JACK STEVENS It was a shame about Gordon, but if it weren't for Tom, he would still be in Yellowknife. Gordon was just out of school and became a stooge for a tinhorn gambler. He and this guy came over to the island where Pete Racine was building a stopping place, and they horned in on it, doing a little bootlegging on the side. The tent eventually became a log cabin hotel business where they served meals. When the gambler split, Gordon stayed with Pete.

Gordon had got in on the staking because of Tom, and the understanding was that he got paid in shares, not money. Afterwards he left for Edmonton and Lieberman bought him out for $1400.

The money looked pretty good to a nineteen-year old, and the situation was such that he had some things to square up. Through some bad business deals, he had once sold his mother's house out from under her for $600, and he'd been in hiding ever since. That was the whole reason he'd gone up north. The lawyer finally bailed him out, and he later returned to Yellowknife 'cleaned but cleared.'

ALICE PAYNE The partners produced a prospectus, obtained a broker's license from the Board of Utilities by posting a $2000 bond, and purchased a miner's license for fifty dollars. It took three months to complete the public share offering of 100,000 shares, which was soon oversubscribed. The only customers missing on the list were eastern mining men and investors, who, in the words of one Cominco employee, "did not think much of gold veins outcropping in the Yellowknife area."

In contrast, many northerners who knew and trusted Tom rushed to sign up. First was Slim Semmler, a trader from Aklavik who worked in opposition to the Hudson's Bay Company and whose wife had a reputation of shooting fifty muskrat before breakfast. Then came Bishop Breynat of the Roman Catholic Church, who lived in Fort Smith. Breynat was associated with Father Gathy, the renowned Oblate missionary. Jim Darwish from Rae, whose wife had nursed Tom throughout the winter of 1934, bought some shares along with the engineer of the *Distributor*, and Kingsley Langfeldt, the RCMP officer, who had shut down Tom's grub pile at Great Bear Lake. Many other northerners came from Waterways, Fort McMurray, and Fort Smith, and included several bush pilots.

In Edmonton, doctors, accountants, bankers and businessmen all bought shares. So did Harry Cohen, of the Army and Navy surplus store, who created a charge account for Tom to buy clothes when he was broke. The legal community was well represented by the H.R. Milner family, George Steer, and a law firm called Friedman and Newson. James Richardson and Sons, a brokerage firm, represented buyers as far away as Montreal.

The list of original shareholders also included many of Tom's relatives by marriage in England: Jim Pawsey bought shares for Rosemary and all of his children. Leonie, Tom's old girlfriend, subscribed for a large amount, along with other family friends.

Ralph Keeping I invested what to me was a princely sum of £25, along with my dear old friend, Mike Martyn. We solemnly sat down after dinner and wrote out the cheques payable to Mr. Lieberman. I must admit that 'gold' frightened us, but for the knowledge that my uncle, Jim Pawsey, and father knew all about Tom's dangerous and exciting prospecting. He had earned their confidence.

David Payne As a child, I knew about the gold mine, and we had some nuggets set around a bowl of cacti on the sitting room table. Tom also gave a talk on the BBC about living with the Inuit when he came to visit.

Alice Payne Dad must have been his own best salesman. His pockets were full of gold samples, and his enthusiasm was contagious. The prospectus contained all the known information. The main vein, which could be traced for 1000 feet before it disappeared into the muskeg, varied from one to six feet wide, and the gold-bearing quartz returned spectacular assay results: one specimen of high-grade ore returned an assay of 60.85 ounces per ton, or $2130. In fact, all of the samples were rich.

The secret that would give the mining property a long life were in the words, "The underground results secured by Consolidated on their adjoining claims ... and the general structure of the main vein in our property, give every reason to believe that the high values will continue to depth."

By April there was enough money in the bank to send Dad north again. He needed to begin development work, and to determine what further costs would be required to bring the property into production. He began these efforts in high spirits, but didn't last the summer.

Tom Payne I set out with Ole Hagen, a green geologist called Tom Hansen, and a couple of other men to draw a map, finish the trenching, and start diamond drilling. This was about a month before break-up. One day I hoisted a large log on my back to place in front of the cook shack door. I was making my way down a billy goat trail that wound around a large pillow of rock. Suddenly the snow on the south side, facing the sun, started to slide. Down I went, log and all, and my left snowshoe got caught on a bit of schist. I couldn't ditch the log, as it would have 'dead fallen' me, so I had to hang on. When at last I spread-eagled, that old hernia that had been bothering me for years burst open.

My buddies dog teamed me up to one of the last planes to leave Yellowknife before breakup. They deposited me at Fort Smith, at Joe Lanouette's hotel, while the pilot left to take another load of freight back to Yellowknife. On his return, he learned that another plane had ripped off a tailskid the day before, so he decided not to chance another trip.

So instead of getting out, I had to spend time in Fort Smith where the doctor was afraid to operate. I suffered the tortures of the damned for four weeks. I whittled out a bat of wood to fit the gap in my abdomen, and held my guts in place with a caribou thong eight inches wide. I drank nothing but powdered milk, orange juice, and beef soup, and every time I coughed or sneezed, I thought I was done for.

At last 'Wop' May arrived on floats and flew me to Edmonton's Royal Alexandra Hospital. Dr. Baker insisted I was not to smoke or drink. As I was a big husky man, he scheduled me for surgery the next morning. I remember they gave me a shot in the arm that made me bold as brass on the way up to the operating room. What was in that hypo I'll never know. The nurses strapped me down, hands and feet, and I remember complaining that the straps were not tight enough and pulling one hand free to show them. Then they told me to take a deep breath, pins and needles hit my throat, and I was gone.

Dr. John McGowan operated on my gangrenous hernia. He was a good knife man, and for about ten days, I got along pretty well. Then one day I got up, went for a wash, came back and sat on my bed, and collapsed. It turned out that I had a blood clot, which had settled in my right lung. And if anyone ever tells you that a blood clot is a small thing, don't believe it!

Just when I was getting over this, phlebitis set in, first on one side, then the other. Seven times I got up and caved in, and finally had to have ten days of hot compresses. Big flannel wool blankets were wrung out in hot water, then wrapped over my legs, and tied on with some plastic sheeting. They had to be changed every half-hour, and it took six days for my temperature and pulse to abate. But I still didn't recover.

My weight had gone down from 230 to 130 pounds, and my joints were swelling out of proportion. No one knew why. Then in walked Dr. Baker, making rounds with an older doctor from Rocky Mountain House. He had chin whiskers, and was dressed in black. Baker pulled down the sheets and said, "Take a look at that!" The old boy snapped his eyes and said, "Why, that man has scurvy!" That did it! This doctor had fought in the Boer War and had been a shipyard medic when most sailors were eating too much salt pork. He had seen lots of scurvy, but no one in Bakers' clinic had that kind of experience.

DR. ABRAHAM HURTIG Throughout all of this, Tom always managed to maintain his sense of humor. We interns often visited patients who seemed to be staying forever. One day, he was lying there in bed, with tubes coming out of him all over, and nurse Aggie Lee walked in with another intravenous line. After she set it up, and had started the drip, she turned to go but Tom stopped her, saying, "Hey! Bring me another one of those! The doctor wants to stay for tea!" Now every time I see someone with an intravenous, I think of him.

TOM PAYNE At the Royal Alexandra, they used to have a cold storage room down at the end of the corridor for patients who died on the ward, or looked like they might before the night was over. Once I wound up in the cold room with just a sheet over me. I wasn't dead yet, though I damn near froze to death. I would have gotten up and walked out, if I could have, but I was far too weak. In the morning, when Aggie Lee came in to pull the blinds, I asked her to wheel me out. She jumped about a foot high, and asked what on earth I was doing in there. So I told her I was waiting for my breakfast! Of course, I could no more eat than walk.

ALICE PAYNE While Dad spent the summer in hospital, Ole Hagen and Tom Hansen worked away with their crew on the veins of the P&G group. Their efforts cost the company about $20,000 for a top notch camp: diamond drilling, pay rolls, transportation, mine tools, hardware, gasoline, explosives, office supplies, and lots of food. $30,000 remained in the company's treasury for future work. By Dad's standards, this place would have been absolute luxury.

Next door, Cominco was also busy diamond drilling and trenching their prospects. In late summer, Mike Finland got sick and Bill Jewitt asked Bob Armstrong to assess the situation. The discovery looked good, but the season was running short. Cominco had decided to sink a little prospect shaft by handsteeling on the Con 10 shear, where gold had been found to a depth of fifty feet.

But only a small tonnage of ore was proven by season's end, and they had to decide whether or not to order mining and drilling equipment for the following spring. Any delay could postpone development for a whole year, but the wrong decision would be expensive. So Cominco sent their manager of mines, their assistant manager, and their head of drilling to Yellowknife to make the decision.

Bob Armstrong I tried to show them around the area, to examine the whole operation and settle the details of future campsites. But these eminent men got out on the dumps, or the piles of rock chips left over from when the shaft was sunk in the ore zone. They wouldn't leave. It's funny how gold affects people. Everyone was excited and shouting, "Look at this one," and I just couldn't pry them away.

Alice Payne It was a good thing for Dad that they stayed there. The narrow high-grade veins cutting through the green volcanics, or buried in the muskeg, did not look impressive. But the chunks of blue quartz laced with gold, littering the dump, caught everyone's attention. W.M. Archibald, the manager of mines, decided then and there to make a mine on the Yellowknife claims. The mill was ordered, a vertical shaft was started near the prospect shaft, and mill construction began in September.

Leo Telfer Bill McDonald, who was in charge of exploration for Cominco, showed me the vein on the P&G claims and the assays, but we didn't expose the vein any further as negotiations were in progress to acquire that group for Cominco. The Con mine's main showing at that time was the No. 10 vein, which appeared to have a limited tonnage of high-grade gold ore. Since Archibald had authorized a one hundred ton mill, the acquisition of this adjoining property became essential to increase the tonnage possibilities. This mill was a very speculative venture at the time.

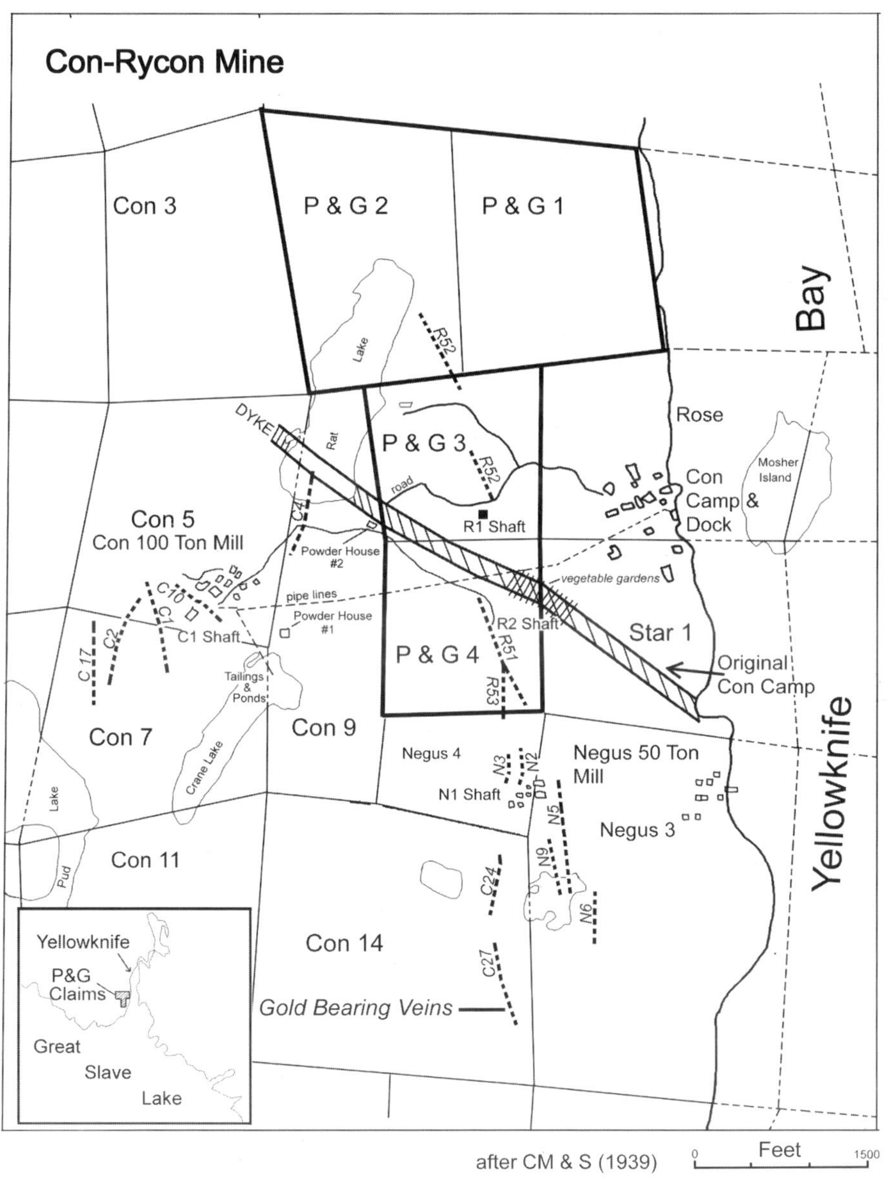

Alice Payne Dad's samples from his first prospect pits contained spectacular amounts of gold, and the results of the summer's drilling on the property next door looked just as exciting. In fact, it seemed that the main vein was on the property of Ryan Gold Mines Limited, which had promised to give Cominco first option. Cominco decided to send Archibald and Jewitt to negotiate a deal with Payne, Ryan, and Wilson, and their lawyer, Harry Friedman.

Tom Payne In late August, just before my last bad spell in hospital, in roared Mickey Ryan: "Archibald and Jewitt are in town to make the first offer to buy the mine. What shall we sell out for?" I told him, "$500,000 cash and a forty percent interest. Not a penny less. Promise me faithfully that if they don't agree, you will say that was your best offer, as we promised them first chance. And then tell them that you now feel free, if the offer is not acceptable, to go ahead and deal with other people. Then leave." Mickey said, "You're crazy, they will never pay that," and stormed out! So I got Aggie Lee to call him back. I told him I was on my last legs, and begged him to do exactly as I asked. Eventually, he agreed.

Later that night I heard the news. Archibald had offered him $159,000 and thirty percent or something like that, and true to his promise, Mickey walked out. Earlier, I had written to R.E. Phelan of

*Yellowknife settlement from the south, Courtesy Jolliffe, GSC.*

Hudson Bay Mining and Smelting, to whom I had once given a barrel of Russian caviar. Andy Dorfman of Mining Corporation had learned of the news through my old English friend, George Keeping, one of the signing officers for the Bank of England and the President of the Investor's Bank in England. They had gathered as a group in the dining room of the Macdonald Hotel!

Archibald recognized them right away, and sensed the competition that Cominco was up against. He sprang into action, called Mickey and Friedman back, and agreed to their offer. And right then and there, they decided how the cash was to be paid: $300,000 when the agreement was signed, and $200,000 in twelve month's time. Ryan Gold Mines Limited would retain a forty percent interest in a new million-share company, and Cominco was to be repaid the $500,000 out of the first profits of the new company, along with any costs incurred through production. The new company was to be called Rycon Mines Limited.

BILL JEWITT As a relatively young engineer, I remember being somewhat shocked when Archibald said he was willing to pay half a million bucks for a sixty percent interest in the property. About as surprized as I'd been when he announced we would put a mill on it, having at the time almost no ore reserves. However, we were lucky.

*Headframe, hoist and mill at the Con Mine. Photo by R.E. Folinsbee*

Mickey Ryan did most of the talking, and Harry Friedman wrote the technicalities. I was staggered. I'd never been in on any deals like this before. Of course, it was Archibald, the manager of mines, who clinched it.

Alice Payne  Reports of the big deal made front page news in the *Edmonton Bulletin* on Friday, September 3, 1937: "Ryan Sells $500,000 Mine," and the following day, the *Edmonton Journal* carried the story, "Faith in Yellowknife Area Pays Richly as Claims Sold." *The Northern Miner* reported that control of Ryan Gold Mines had been sold, but didn't print the details. They were later officially released and appeared in a January issue under the headline, "Ryan Bros. Make Fine Deal," with a history of the operation and a photograph of Cominco's plant at Yellowknife.

This was the first gold mill in the Northwest Territories, complete with steam heat, and the mine came into production in 1938. The deal for the four P&G claims was one of the biggest sales of its kind in Canadian mining history. Dad, still flat on his back in hospital, told the doctors that he would have no trouble paying his bill.

Tom Payne  Eventually I recovered by eating mostly potato peelings, for the ascorbic acid. What a diet. The doctors kept me in bed for thirty days, and I made up my mind that if I didn't survive this time, nothing else would matter. It took two orderlies, plus Aggie Lee, to hold me together. Finally I got the strength to get to the door, via the bed and the wheelchair, then with crutches and later a cane. Soon I was allowed out by taxi from the Emergency Entrance, to visit Calhoun, owner of the King Edward Hotel, the local watering hole for all the northerners. I was back in business.

# Romance and Roses

## Chapter 8

Tom Payne After Harry Friedman drew up the legal agreement with Cominco, and I had signed it, Dr. Baker insisted that I go to Nassau for the winter. He told me, "You'll die for sure if you stay up here, and you must have heat, sun and proper food." I was more than happy to take his advice; the only trouble was that I couldn't persuade one of the nurses to go with me.

I first met Olga at a New Year's party at Ned Chisholm's house in Edmonton. All the Northerners were there: prospectors, bush pilots, the Ryans, everybody. We had the best time. Olga had come with Steve Parlee, a doctor who once worked on the portage at Fort Smith, slinging boxes to pay his way through medical school. I went back up to Yellowknife after the party, to work on the claims, but I couldn't get her green eyes and soft voice out of my head. I made up my mind to track her down, but when I landed up in hospital with my hernia, there she was, the assistant night superintendent.

Olga Payne My job was to look after all the public wards. I always took over when the head nurse, who did all the private and semi-private wards, went on holiday. I went on duty at seven p.m., and worked a twelve-hour shift. I also was the scrub nurse for the operating room in an emergency: fractures, acute appendix, whatever came in.

Over July and August, I was doing my rounds on the wards and Tom always wanted to talk and tell stories, but I just didn't have time to sit and listen. I'd try to get out of there so fast. I must have looked pretty good at night. If it had been daytime, he probably wouldn't have noticed me.

After Tom's health improved, he was allowed to go out during the day and return to the hospital at night. He used to date some of the other nurses, and everyone thought they had him on a string. But whenever I had my nights off, he would invite me to dinner at the Macdonald Hotel. And large bunches of beautiful red roses would appear on my desk in the daytime.

He sent those flowers to me nearly every week, and one of the doctors remarked, "There's only one man in town who can afford those, and that's Tom Payne." So everyone came to look and I'd be pleased they were mine, but I pretended not to know who had sent them. He also gave me a beautiful silver fox, and everyone wondered where I got it! He wanted to buy me even more things, but I wouldn't let him. I didn't think it was proper.

As nurses, we'd all been warned to be professional. The administration was very strict: do what you're told, and a patient is just a patient. We weren't to take on their troubles, or listen to their life stories. Just do the treatments and give the pills, with no emotional involvement. I was a head nurse, so I had to set a good example for all the others.

ALICE PAYNE Mother was a farm girl from Forestburg, Alberta, southeast of Edmonton. Until she turned ninety, she would never reveal her real age, but actually, she was born on December 13, 1907. She was the third in a family of ten children, of whom eight survived into adulthood. A set of twins had died of measles as infants. She was christened Olive Amelia Banks, or Olga for short. Everyone in her family was encouraged to get an education and become a teacher. When she graduated from high school, Mom spent a year in Camrose at Normal School, and then took a job near the Saskatchewan border.

OLGA PAYNE It was easy to move locations back then. I could cram all of my clothes and possessions into one suitcase. I had my lingerie, and one navy dress with little buttons down the front, just what you would expect a schoolteacher to wear. I boarded with a nice family, but was terribly homesick. After a year, I returned to Forestburg with enough experience to teach grades eight and nine. I gave my parents $25 every month for rent, and saved enough money to pay off their farm debt of $400.

As I settled into my new job, a farmer who lived a few miles away had picked me out to marry his son. He proposed this to my dad and tried to make a deal, because that's the way it was done in Romania. Maybe he thought our family were foreigners too, because everyone called me Olga. Anyway, he produced a tape measure, wound it around my calf, and pronounced me acceptable: "Big. Fat. Strong like hell. Good wife for Sam." All that was true. I was hardened by the farm work, getting up at five every morning to milk the cows, but I was definitely not interested in Sam.

Teaching didn't agree with me either. Since I was fourteen, I had dreamed of becoming a nurse. So during the summer I went to the Royal Alexandra Hospital, in Edmonton, to see Miss Munroe, the Superintendent of Nursing. When I inquired about training, she said that applying few weeks before the fall term was not the ideal time. I felt very discouraged. So we talked a little more, and as I walked away, she called me back, and said, "Miss Banks, if you would like to start in September, you may submit an application." So off I went, bought some pink material, and started to make my probationer's dress. You had to make your own in those days.

Nursing training took three years. We had room and board, and the first year, a salary of $350. That amount almost doubled in our final year. My friends and I were so short of money that we used to walk downtown on our half day off, just to save a few coins. We frequented a little coffee shop across the street from the hospital, called the Malta. That's where I learned to smoke; it was the cool thing to do.

After graduation, most of the class left to find work. Every year, the hospital hired a couple of nurses from the new crop, and in 1934, I was one of them. So I stayed on, and became Assistant Night Superintendent. Miss Munroe told me to keep my ears open and my mouth shut, and it was good advice. I loved my work.

Nurses always seem to be hungry, and many of the interns would ask me out to dinner. These young doctors were so different from the farm boys. When I was seventeen, a fellow from a nearby town wanted to marry me. He used to play baseball with my brothers. His family was going to move to the States, and he wanted me to go along as his wife. I was fond of him, but I wouldn't go. He gave me a green set of brushes and combs to put on my dresser so I would remember him. I still have the mirror.

TOM PAYNE I had also made the wrong pitch when I asked Olga to come to Nassau with me. I told her that three months in the Bahamas would do her good, and it wouldn't even cost her anything. She was very insulted and asked me to consider hiring a nurse when I got there! I should have realized that red roses and some Dewar's scotch whisky were in better taste. I had to take my five grand and go south alone.

So I wrote to her from Toronto, where I had a room at the Royal York Hotel. The dining room had a dome-shaped, hand-painted ceiling and I ordered fresh lobster for dinner. I asked her not to condemn me, or be embittered, and to remember that she had seen the best and worst sides of me. She upset me more than any other woman, creating such intense and violent desires that I wanted her desperately. So I complained that I was alone in a lovely place, and begged her to write back.

My next stop was Montreal, where I visited John Molson, the brewer, who was an old school mate. In his private plane, we flew out to a beautiful spot in the Laurentians for the weekend, where I again wrote to Olga saying that I wished she were there, and that I had never missed anyone so much in my life. Still in recovery, I was just beginning to realize that it had taken seven months to get in this mess and it might take another seven to get out of it.

Finally, I arrived in Nassau. I spent the loveliest four months that one man ever had at the Royal Victoria Hotel. Perfect weather, white sandy beaches, and gin and tonics put the life back into me. At first, it was all I could do to paddle out and let the waves bring me in. Then after a few weeks, with the healing effects of the salt water, I could swim quite well, and started having a whale of a time. I threw away the crutches, then the cane, and got around under my own steam, all browned up.

One person I particularly enjoyed was Harry Oakes, who had made his fortune in the mining camps of Ontario. We went yachting and swapped stories of our prospecting adventures - the disappointments and the triumphs. The Nassau crowd made celebrities of us, but in private, we spent our time reminiscing about the years of struggle that brought us our fortunes.

Harry had abandoned medical school for prospecting, and after working in various frontier mining camps, staked the Lakeshore and Tough-Oakes property near Swastika, Ontario in 1912. He became immensely wealthy from the former, but ill health, failed political dreams, and rising taxes prompted his move to the Bahamas in 1935. Besides being a real estate developer, he was elected to the legislature. His philanthropy led the King to give him a Baron's title. The sad twist is that one night, several years later, he was murdered in his own home. That crime remains unsolved.

When Olga's first letter arrived I was overjoyed. Thoughts of her were still fresh in my mind in spite of all the diversions. Things really happened in Nassau, people knew how to enjoy themselves. They drank and raised hell all night, and lay in the sun all day. What a life! It gave me food for thought. However, I began to get bored, and wrote to Olga that I would be home soon, and to prepare herself for a wedding ring.

OLGA PAYNE When Tom came home in May, he invited me to go to Yellowknife with him. But I wasn't convinced that he wanted to get married, neither was I going to quit my nursing job on a whim. We had some anxious moments, especially when Dr. Baker forbade the trip, and Tom went anyway, in spite of the medical opinion. His answer was that it was cheaper to die there than in Edmonton. So we continued our courtship by mail, and every week the plane brought new letters. He asked me to buy a magnifying glass for him, and he kept thinking up repairs for me to do with his car. Nearly every letter asked about what I was doing. He begged me to stay out of trouble, especially on holidays. That sure made me smile, as usually I only went to Forestburg.

During these months, I discovered that he could be quite a rascal. One day he was in Vic Ingraham's hotel lobby in Yellowknife when an Edmonton store manager appeared with his favourite sales lady. They had quite a time, and everyone was wise to what was going on. The woman complained to Vic about her door squeaking, and of course, Tom said, "How's your bed?" creating a roar of laughter from people within earshot. "She's as mad as hell at me," he wrote, "It kind of hit a soft spot."

Tom tried to take it easy. He wrote that he could only "crawl around and lie on the rocks," but did regain his strength. When out in the bush, he paddled along the lakeshore and let his two hired men do the heavy hiking. According to Tom, some prospectors had made tremendous finds east of town. Planes buzzed around like wasps, and people spent lots of money. He said the majority of new claims were pretty remote, and had to be very good to turn a profit. But by late August someone had produced a chunk of rock that contained $7500 worth of gold. Measuring one foot square, it was a regular gold brick. This started a stampede, and Tom was in up to his eyebrows. He even took out a prospecting license in my name, and a bunch of others for various family members.

From his letters, I could imagine him enjoying the peace and quiet of his camp, located on a lake north of the Thompson find, which had a very rich show. There were loons, ducks, and fish galore, with trout jumping in the lake outside his tent. Early in the morning before the daily hum of planes, he would gaze over the lake and smoke a cigarette. When his two rock hounds were out in the bush, he wrote to me saying, "Too bad you aren't here too." He didn't sound like the typical hard-bitten prospector.

By mid-September he wrote a long rambling letter and finished by saying, "Only ten days to go." He felt that the summer's work didn't amount to much, but he was happy for the hunt and had acquired a good tan. He said that he'd never missed anyone so much in his life. Sometimes he got very sentimental and told me things he would never say in person, so I came to understand him better. He believed himself to be a peculiar person, without many friends, and there were darn few people he really trusted.

Our good times together, and the sheer pleasure of each other's company, astounded him. It was almost a confession, as if he might never see me again: "Olga Dearest, I'll never forget you and happen what may, you will always remain very very dear to me, someone very lovely and possessed with great understanding. You have been all that's good to me and I can't forget it. Foolish, I know, but it can't be helped. I miss you a lot, and it won't be long until I'm back, and as you say, it should be good for a couple of beers." By the time he came out of the bush, he was ready to make a commitment. After all, it was his third time of asking!

I was starting to get very lonesome too, and was relieved to get his wire from Yellowknife on September 16th: OLGA BANKS ROYAL ALEXANDRA HOSPITAL EDMONTON ARRIVE COOKING LAKE LATE CANADIAN AIRWAYS FIVE THIRTY PM IF POSSIBLE MEET ME PAYNE.

He wanted to get married right away, but I said that the hospital needed a month's notice, and I couldn't just pick up and leave. I didn't get far with that argument. With more red roses, Tom charmed Miss Munroe into allowing me to retire honorably. In those days, if you married, you either quit or were fired. As a nurse, all your attention had to be focused on your patients, not on husbands or families.

We headed for Calgary and Tom took me to Birks. He bought me the most beautiful wedding ring I'd ever seen. It was yellow and white gold with five diamonds, and to go with it, he picked out a very classy wristwatch. Then we stopped at the government liquor store to buy a few bottles of Mumm's champagne.

TOM PAYNE We got married in Vulcan, southeast of Calgary, at nine in the evening on September 24th, 1938. It was a small wedding. Olga's younger brother, Beverly, and his wife Ethel, stood up for us as best man and matron of honour. We had one hell of a time flushing out the parson, a Reverend Dobson, and I gave him a twenty-dollar bill for his trouble. The poor man complained that he couldn't make change, and it took a while to persuade him to keep the bonus! The next day we left Bev's house in a green Dodge coupe that I had bought and we headed for California.

OLGA PAYNE We went through Great Falls, Montana, and visited a friend of Tom's from Felsted. He introduced us to our first American bar. I drank a whisky sour and Tom drank Planter's punch, and there was a poker game going on downstairs. Usually there were bars for men where no women were allowed, and bars for ladies and escorts, with no singles allowed. And absolutely no gambling! In the States, once they got rid of prohibition, there were no restrictions. It was all wide open. We toured the Anaconda Copper Refinery, and stopped in Butte. What a town! Their own description read "A hootin' tootin' roarin' hell-raising town that drinks its liquor straight." Quite impressive.

In Salt Lake City, I spent three days in hospital with suspected appendicitis, but after a complete overhaul, the doctor released me. We drove through the great salt desert, all white and dazzling in the sunshine. We checked out all the old mining towns, and met some Nevada cowboys, with their big ten-gallon hats and bowlegs.

There was more excitement when we got to Sacramento. Our hotel lobby was full of women wearing fur coats (only rabbit fur!) and men in lounge suits, and they turned out to be the MGM crowd. Mickey Rooney was there, and I had a good stare. Heady stuff for a farm girl! We finally arrived in San Francisco and registered at the Sir Francis Drake Hotel. We dressed up for dinner, and Tom looked swell. Cocktails, dinner, music, I could hardly believe it was me right there. Wonderful.

We drove home along the coast road through the giant redwoods, and stopped at all the viewpoints to look at the ocean. The trees and plants were luxurious, it was all so beautiful: vineyards, orchards, flowers, everything. We were away about three weeks, but I would remember our trip forever. At the time, I didn't know we would be honeymooning for a whole year.

Tom Payne More money had come in from Cominco regarding the Rycon mine, and I had a yearning to see England again. So our next stop was the travel agent. I booked us on the *Cunard White Star Line* for London. When we arrived in New York, the first thing I did was take Olga on a shopping spree. Our ship was called the *Georgic*, and we boarded just before her birthday in December.

Olga Payne Tom was very generous. He had impeccable taste, and wanted me to look good. I had my Mackenzie wild mink coat but not much else other than my nurse's uniforms. We spent the day at Altman's, and bought a whole new wardrobe. I had to have a wool suit, a little black dress, two evening gowns (one light, one dark) and an opera wrap for evenings. The wrap was made out of black silk velvet, with a white fox collar. I looked spectacular. Tom even bought me a new suitcase and made me throw away my old one. He also insisted that I cut out the labels in all my new things except the ones from Altman's. He said the maids would look.

We had a wonderful voyage. I loved watching the ocean, even when it was gray and stormy. The bow of the ship would rise and fall into the big waves, and it didn't affect me at all. I enjoyed every single minute of it, and hated to leave the ship when we docked at Liverpool. I had absolutely no idea what lay ahead.

We took the train to London and out of the window, I saw the brick houses, with their chimney pots by the thousands. When the mist cleared, the countryside looked very pleasing. The farms looked like gardens divided by hedges! When we arrived in the city, it was snowing and blowing. I almost froze, as the Piccadilly Hotel did not have central heating.

We had arranged to dine with John Gardner and his wife. It was then that I found out Tom hadn't told anyone that we were married, planning a big surprise! However, things backfired when the Gardners invited Leonie Gray for dinner. What a shock! She was all excited, dressed up, and naturally assumed that Tom had come back to get engaged. From all accounts, he'd considered Leonie as just a friend, and never had any real intentions. As I sat there in my New York clothes, I tried to be polite. In truth, I was mortified and felt like clobbering him. It was a hell of a night.

The next day, Tom was raring to go. I felt like going home, but after breakfast in bed, things improved. The truth was that he was happy to be back in England, and wanted to buy himself some new clothes at Moss Bros. I was used to seeing him in his old prospecting duds, which were nothing like the tweed plus-fours and yellow knee socks he wore to lunch that day. Relaxing in his tuxedo at night, he looked like the country squire he had always intended to be.

We took the train down to Colchester, and arrived at J.A. Pawsey's, who had invited Bill and Rosemary, and a few others. They were all so grand, and this eased the painful experience. We had a pleasant evening with good drinks, good food, and a proper night's rest. Many people dropped in over the Christmas season. We went to see Tom's sister at Witham, and drove down to Clacton to visit Annie, who was overwhelmed to see him. He absolutely refused to see his father, Frank, who would live another eight years.

Tom had a lot of 'men only' outings. He and Bill went yachting, and attended a dinner at the Colchester and East Essex Cricket Club. I was sorry to have missed it. There were ten courses for dinner. After the hors d'oeuvres, there was consommé, fried Dover sole, pheasant savaroff with cream of sprouts, then a kirsch sorbet to cleanse the palate. Next came the roast saddle of mutton with red current jelly, and potatoes, French beans, and braised celery. They finished it all off with iced chestnut pudding, mushrooms on toast, and coffee. Hard to imagine a menu like that.

We visited the Keepings, and had dinner with Mont Everard, who lived at the Coggeshall Abbey. The place had been built in 1100 A.D., and had beautiful woodwork, aromatic with age. Everything in England seemed ancient to me. One night, Bill and Rosemary served us Stilton cheeses with some of Dr. Payne's port, bottled in 1896. It was so old and mellow that I never forgot it. Later, Tom used to buy Stilton cheeses at the market and give it to all his friends in Edmonton for Christmas. However, we never drank port wine quite like that again.

Before sailing home, we stayed at the Regent Palace in London and took in some stage plays. We browsed in John Gardner's antique shop near the gates of Buckingham Palace. John's dad had been an auctioneer, and now he looked after the shop. I was just beginning to like London when Tom decided it was time to leave.

TOM PAYNE Soon all we heard about was the war. I began to think we might never get back to Canada. So in case all the English ships were commandeered overnight, we changed our booking from the Cunard line to an American one. We left on the *President Harding*, an old tub to be sure. The captain, who wore a cloth cap tilted to one side, invited us to sit at his table, but I politely refused. We just didn't feel comfortable with that crew, and I didn't feel like buying them drinks.

It was an awful voyage and there were forty-foot waves most of the time. One night, I was chucked out of bed. Olga teased me saying that I had made a beautiful four-point landing, but after that we tied ourselves into our bunks and into our chairs at mealtimes. The ship was crowded with Jewish refugees heading for New York, and the passengers didn't feel much like socializing. We were glad to reach shore in one piece. When we got home, we got quite a jolt. At the Capitol Movie Theatre, on the newsreel, there was our captain, with his peaked cap over one ear, claiming that the stormy seas had made it the worst trip of his career.

OLGA PAYNE Back at Edmonton's King Edward Hotel, I was starting to getting bored. Our lives consisted of entertaining all the Northerners, going to movies, riding in the car, and studying the real estate market. My life was nothing like I'd had in the emergency ward. So when Tom suggested we go to Yellowknife, I was all for it. He went first, to get the camp ready, and I followed after break-up. The train went as far as Waterways, then I went by taxi to Fort McMurray. I was all dressed up in my New York clothes and everybody stared at me in my stiletto heels. Once they found out I belonged to Tom Payne, I was treated like royalty.

From Fort Smith I caught a pontoon-equipped airplane to Fort Chipewyan. During the six-hour layover, I dropped in on Ned Chisholm, whose parents owned a hotel there. Then we flew through the smoky twilight to Yellowknife. I didn't like night trips, but North Sawle, the pilot said, "That's all right, we will soon be flying into daylight." The plane was very primitive, just one propeller buzzing away and no seats, only cargo. I sat on a crate of oranges. We flew at about 500 feet, following the edge of the river, and you could look down on a string of forest fires. There were five other passengers including a priest, an author, and a woman with a wild looking hair-do. She was none other than Vicki Lepine, who ran both the laundry and the red-light district in Yellowknife's Peace River flats!

When I arrived at the dock, just below Harry Weaver's trading post, Tom was there to meet me. It was almost at midnight but still light, and he half carried me up the hill to Vic Ingraham's hotel. The next day I bought some bush clothes. It was New York in reverse. Vic's Hotel was a real jamboree. All the prospectors were there. It had thirty-two rooms, and the whole place was jacked up on stilts. From every window, you could see the drunks 'sleeping it off' on the rocks.

Tom sure started something at Yellowknife. The main town site was like a large construction camp. Buildings were going up everywhere. One night Bill McDonald, the geologist, and his wife invited us to their house at the Con mine for dinner. The mining camp had been modernized in an amazingly short time. Cominco had small houses for families, a bunkhouse for the single men, and a huge dining room that ran around the clock. When work was over, everyone met in the recreation hall, and if you were sick, Dr. Oliver 'Ollie' Stanton

*West shoreline of the town of Yellowknife, 1939. Buildings include drugstore, Vicki Lepine's laundry, the Corona Inn, and Weaver and Devore's trading post. Photo by R.E. Folinsbee.*

treated you in the company hospital. Burns and Company had a meat freezing plant down by the dock. The whole community was heated by steam, with the pipes insulated in thermidors, which ran over the rock outcrops and doubled as wooden sidewalks.

Our prospecting plans were delayed by a series of terrible forest fires. Sometimes the smoke was so thick that you couldn't see the water. Then one day we saw the flames rise over the hill at Ptarmigan, across the bay. The firefighters were trying to save their camp with water pumps and men working steadily. Finally it rained, and the sky cleared enough for us to take off.

The first camp was at Pensive Lake. Tom was determined to travel in style, and it took three canoes to hold all our groceries and gear. He had several new white cotton tents with no holes, a good cook stove, army cots to sleep in, a canvas bath tub for washing, plenty of food,

*Tom and Olga Payne on the porch of Vic Ingraham's hotel, Yellowknife, 1939. Photo by R.E. Folinsbee.*

and me to make bannock and keep him company. He even had a radio set in case of disaster. He always picked campsites out on a point, with a view of the lake, and a breeze to blow away the mosquitoes and black flies.

Most days were sunny and warm, and we often paddled around the lake. Or, Tom would go off with his two prospectors, Pete Lauder and Pete Davidson, and his hand-steel men, Phil Friel and Alex Mitchell. I stayed around camp fishing, cooking, or washing clothes. When it rained, I took cover and crocheted my tablecloth. By the end of the season, I had fifty-six patterns to sew together.

Occasionally we had visitors, or the pilots would drop in and I would make ham sandwiches and coffee. Sometimes we just served the Hudson's Bay overproof rum. There were so many camps around us that I wasn't lonesome at all: the Geological Survey, Cominco men, Ontario prospectors, and even the dentist from Yellowknife. I took lots of photos with my box camera, and Brownie, our dog, was my constant companion. He even saved my life once.

*The back view of Ingraham's hotel, with discarded liquor boxes in foreground. Photo by R.E. Folinsbee.*

It happened one day when I wandered away from camp without taking a compass or map. Harry Ingraham, Vic's brother, had been giving me some prospecting lessons. He showed me what vein gold looked like, and taught me how to crush the rock with a pestle and mortar for panning. At first I found nothing, but then I found a little gold while eating lunch on a rock. When I took the sample back to camp, it caused a minor rush and I was very pleased with myself. After that, I went out often.

*Olga Payne with Brownie, the dog that saved her life at Pensive Lake. Payne family collection.*

That particular day, things looked familiar, but about four o'clock, when it was time to return, the trail seemed mixed up. All the jack pines looked identical and I was terrified. I sat down to consider my options, and then I had an idea. Thank goodness Brownie was with me. I jumped up and yelled, "Bad Dog" and "Go home." I kept on shouting and stamping my feet, and the poor dog was completely puzzled. He tried to whimper and cuddle against me, but finally I picked up a stick and chased him away. Then I followed him, and he led me back to camp where I burst into tears, and hugged his furry neck. Tom didn't know whether to be mad or glad, but Brownie was delighted to be forgiven. After that, my interest in prospecting evaporated somewhat.

In all, we'd pitched four camps, the last being at Desperation Lake. By mid-September, the snow began to fall, and it was time to go out. Tom had made a number of finds, so with plans to return, we cached the gear in a tree for the winter: tents, canoes, stove and all.

*Tom Payne washing in his luxury camp. Payne family collection.*

*Tom welcoming a Junkers supply plane to Desperation Lake with Alf Caywood and Vic Stevens standing on the pontoon. Payne family collection.*

*Looking for colours at Desperation Lake, a good find that turned out to be poor. Payne family collection.*

*Two handsteel men work on the find at Desperation Lake while Olga Payne looks on. Payne family collection.*

*Making a bear proof cache, Lauder Lake, 1939. Payne family collection.*

Our last meal was a banquet. The lake trout were running in the Cameron River at the north end of Pensive Lake. Tom caught five fish in five casts. We rolled them in flour, fried them in a pan with the last of our butter, and ate them with bannock for dinner. It had been a wonderful summer.

# Interlude

## Chapter 9

Olga Payne After returning from Yellowknife, we moved temporarily back into the King Edward Hotel. Then we found a house on 122nd Street, just off Jasper Avenue. It was a good thing we bought it, because in December, I learned that I was four months pregnant with Alice.

I could see that the money was going fast on travelling, living in hotels, and purchasing a car and furniture, not to mention grubstaking the two Petes, hiring the hand steel men, and flying about in the territories. We already owned an apartment block called the Leamington Mansion, on 114th Street, but I persuaded Tom to invest in another block of suites and half a dozen more houses as revenue properties. It wasn't a hard sell.

While in England, Jim Pawsey kept telling Tom how well he'd done in real estate. Over the years our investments proved to be good ideas too. Tom doubled his money when he sold, and the rents helped us manage financially when prospecting ventures failed and the Yellowknife mine paid no dividends. Jim actually came to visit, and was very impressed with Holgate House, our 1912, fifteen-room mansion on Ada Boulevard in Edmonton's east end.

Alice Payne   We moved into our new home in the winter of 1942. I was about two years old and my sister, Elizabeth, was an infant. It was bitterly cold and windy, and I stood on the back porch with my mother who was trying to soothe my crying sister. We waited for what seemed like hours for moving vans, and for Dad, who finally appeared with the door key.

The former owners had spared no expense. They had imported beautiful chandeliers and large leaded glass windows, giving it a very English feeling. The outside was constructed of red brick, and a spacious verandah wrapped the entire front section. White Grecian pillars supported the upper levels, which had gravel stucco siding. A circular driveway wound through the grounds, connecting a separate building at the back that could be used for a carriage house or servants' quarters.

*Tom with daughter Alice on the front porch of his first house, 1941. Payne family collection.*

Inside was very beautiful. The floors were of polished oak, and the walls were panelled in oak and hand-painted linen. Wide arches opened from the large front hall into the living room with its enormous fireplace. The music room with its Grecian frieze decor on the ceiling and harbour bay window was a lovely setting for the grand piano. Upstairs on the second floor there was a bright, glassed-in conservatory that my mother filled with tropical plants.

The main hall led past the stairs to Dad's den, complete with a massive oak desk, leather chairs and a matching sofa. His bookcases contained prospecting publications and held rock samples of everything from gold to pitchblende. He stashed drill bits, Geiger counters, and sample bags behind the furniture. In the middle of the room stood a three-dimensional glass model of the Rycon mine.

Hobbies long neglected resurfaced, and a certain coldness developed between my parents in the fall when duck decoys littered the floor. Mom was exasperated when she had to clean under the fishing rods and shotguns that were tilted in every corner. The built-in liquor cabinet was stocked with fine wines and spirits, and promised a hearty welcome for guests, except that the den lacked space for anyone to sit down.

*Tom's second house, the Holgate House, on Ada Blvd. in Edmonton. Payne family collection.*

Elizabeth Payne I can still remember the elegance of Holgate House, and the great number of people who visited us from elsewhere. Dad entertained them with Dewar's whisky and tall tales, while we were usually put to bed early.

Alice Payne As children, we were expected to play at the rear of the mansion, but found the servants' staircase dull compared to the magnificent slide provided by the oak banisters that cascaded into the foyer. When banished from there, we ran around the uninhabited third floor, and often staged elaborate theatrical productions.

Outside, Elizabeth and I were often penned up in a triangular area surrounded by a six-foot fence. We tried to dig our way out several times, aided by the family dog, but decided against it after a few of Dad's stories about walls in England with broken bottles along the top. Our property was well treed, with willow, elm and maples. We had a hammock and swings, and a sand pile with a hose and sprinkler for making mud pies. When the playing got rough, Dad would turn the hose on us to cool things down.

As we grew older, bicycles gave us more mobility. We attempted to build a private cave in the riverbank across the road, furnished with boxes and plastic curtains where we could stalk imaginary cowboys and Indians. But after someone shot the heel off my cowboy boot with a BB gun, Mom declared the area out of bounds! I finally convinced my parents to buy a horse, but was puzzled as to why it couldn't live in our back yard.

I spent many summers in Yellowknife with mother's sister, Florence, and her husband, Bill, who lived in the house on Jolliffe Island and managed Dad's prospecting camp. My sister, Elizabeth, was not so lucky. She suffered from a number of allergies and severe asthma, which made it impossible for her to live the sort of uncivilized life that I coveted.

During the winter, we hauled cordwood on our toboggan to burn in one of the four fireplaces. When we complained about the work, Dad flooded our circular driveway for a rink, and we learned to skate by pushing a chair around on the ice. Perhaps this gentle beginning was what made Elizabeth choose skating instead of swimming as therapy for the polio that later threatened to cripple her young life.

*The shack on Jolliffe Island, Yellowknife, with Florence Peters, Alice, Elizabeth and Olga in the foreground. Payne family collection.*

*Tom paddling a canoe near Jolliffe Island in summer with Alice and Elizabeth Payne. Payne family collection.*

My mother was completely occupied in caring for our home and us. We often had a maid to help with housework and prepare meals, but Mom spent most of her days mopping and dusting, waxing and polishing, and sometimes must have wished she had stayed in nursing. Dad always called the hired help 'the slaves,' but she was the real slave.

He loved strolling around his estate and bothering the gardener. The English flowerbed at High House had been quite famous and he insisted on trying to duplicate it. With a total disregard for nature, he insisted on planting things that were sure to shrivel in the dry Alberta air. If the gardener did coax some exotics into bloom, he would be indignant when Dad wondered why they were so measly!

Our lives were a curious mixture of luxury and thrift. We had none of the new electronic gadgets then available, and in spite of the size of our home, a telephone extension was considered non-essential. Even the word 'television' was only one rung up from swearing. We could never decide whether this was an expression of Dad's dislike of things technically modern, or a trace of the stinginess left over from his years of deprivation in the north.

*Alice Payne in front of Holgate House, 1955. Payne family collection*

Olga Payne When the Second World War encroached on our lives, I knitted baby clothes and worked at the Red Cross blood donor clinics. Tom was also determined to make a contribution to the war effort, having missed out the first time. He decided to go prospecting up the Alaska Highway for strategic minerals like silver, tungsten, and gypsum as well as gold. He assembled all the government reports he could find, and packed the slim collection of maps then available for northern Alberta, British Columbia, and the Yukon Territory.

He bought an old International pick-up outfitted with a canvas top, benches for beds, a stove, and enough storage space for samples, extra gas, and equipment. He'd applied for a permit from the Canadian Government, but set off with his prospecting cronies before the documents arrived.

*Tom Payne with a Canol project engineer from New York. Payne family collection.*

*Tom Payne's Alaska Highway truck. Payne family collection.*

TOM PAYNE We got into trouble at the first American checkpoint, and were forbidden to proceed without identification. "Don't forget we're fighting a war here," the American colonel told me, but that didn't impress me. My reply was, "Don't forget we're paying you." I knew that the Canadian Government was footing the bill for part of the highway project, and I could probably talk my way through, but Pete Lauder's situation was different. He was afraid that the guards would still have a warrant for his arrest because of his past. So we drove back a few miles and let Pete out. He circumvented their fences, and we picked him up again on the other side! After a while, the permits caught up with us, and we were no longer *persona non grata.*

Most of the bush road was under construction and driving was pretty rough. We needed a new rear axle at Dawson Creek, and headed for the Watson Lake area to try and contact Dr. Lord of the Geological Survey. His party was the first to do any proper fieldwork in the southeast Yukon, and we aimed to ask him if he had any leads.

We drove on to the Rancheria River, a little stream meandering down a rocky riverbed. The water wasn't deep enough for a canoe, but it looked like a good place to camp. We parked the truck right out in the middle, on a gravel bar, where a good breeze would blow away the bugs. The weather was quite good, and after supper, we went to bed under the truck instead of inside it.

Then in the middle of the night, the ground started to tremble and I could hear a strange rumbling noise. It sounded like a herd of stampeding buffalo. I jumped up, woke the others, and we piled into the truck and got out of there. As we reached the riverbank, all we could see was a solid wall of water rising down the riverbed, destroying everything in its path. As we reached higher ground, uprooted trees washed over the very spot where we'd been sleeping!

We had quite a trip, altogether, but I was shocked to see bulldozers with flat tires abandoned by the side of the road, and enough discarded junk to form a construction company. There were stacks of unused food piled up at depots, and we watched as some soldiers punched two holes in a can of fruit, drank the juice, and threw the container away! They had no idea how to live off the land. Later we taught one war department official the art of baking bannock in an oven made of hot rocks and sand. That made me feel better. After all those years in the north, I couldn't stand to see things wasted.

ALICE PAYNE It would have made little difference if Dad had found fifty gold mines; he would still have lined his shoes with cardboard, and used cigarette boxes for notebooks. Lunch bags were so highly valued that after using one, he carefully folded and then stored it inside his billfold. Reputable witnesses have recounted that after such a performance, he would absent-mindedly leave the wallet behind on a rock.

When it came time to spending money, new prospecting ventures seemed his first priority. Dad's idea of a surprise was a new car or horse, and any day was time for giving Mom diamond rings or a mink stole. But she had to resign herself to going to the best places with a husband dressed in shirts with the elbows missing, and old bush khaki pants held up with a length of binder twine. He made a point of wearing his old school tie until it was inadvertently hooked

*Members of the Prospectors and Developers Association at a meeting in Edmonton, Alberta, Feb. 29, 1944. Members identified are Jack Horan, Tom Haegerty, Tom Cassidy, 'Spud' Arsenault, Neil Campbell, Vic Ingraham, John Michalson, Albert Gagnon, Mr. Molstad, Alex Mitchell, Tom Payne, Jim McAvoy, Viola and George MacMillan, J.D. Nicholson, and Dr. Lilge, Cliff Lord. Courtesy Prospectors and Developers Association/NWT Archives.*

into a rag rug. The one time he could be guaranteed to dress up was at the Prospectors and Developers meeting, which was held in Edmonton in 1944.

Our brother, Tommy, was a late arrival in 1949. Mom was already in her forties and Dad was fifty-eight. As Tommy grew, he took my place painting duck decoys, learned how to run machinery, and became a miniature version of Dad. Because of our differences in age and physical ability, we all came to lead separate lives. I became a geologist, Elizabeth studied fine arts, and Tommy became a short line railway entrepreneur.

After fifteen years, it was no longer physically possible for Mom to handle all the work, and we moved to a smaller, ranch-style bungalow on Saskatchewan drive, near the University of Alberta. The house had a beautiful fieldstone chimney, but inside the doors were hollow and underneath the broadloom carpet the floors were made of plywood. Like Holgate House, it had a beautiful view of the river valley, and we spent many evenings in the sunken living room admiring the western sky.

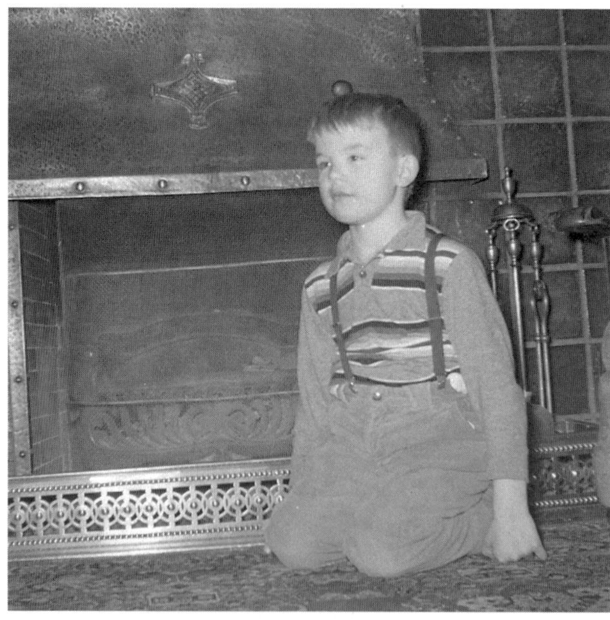

*Holgate House, 1955. 'Tommy' Payne in front of the living room fire place. Payne family collection.*

We had problems selling the big house. Few people wanted such a huge place, and it was prone to attack by vandals. When some kids had a party in the kitchen and then broke a few of the bannister railings, that was the end for Dad. He went over and sat in the dark with his shotgun, and fired a blast at the next gang of marauders. Highly illegal, but definitely effective. I think we eventually sold the house to a chemist with seven children.

# Faith

## Chapter 10

Tom Payne I always banked on finding another mine in Yellowknife, so every year I sent my prospectors out into the bush. My biggest hopes were for Payne Yellowknife Gold Mines Limited, a company formed to develop three groups of claims that we'd staked. On the SP group at Russell Lake, north of Rae, several gold showings had been uncovered along a length of 1500 feet. At the PLD claims, staked by the two Pete's during the summer of 1939, when Olga and I were out in the bush, surface assays ran at $2000 to $3000 per ton and the showing ranged from five to twenty feet wide over three miles. On the west side of Gordon Lake, small showings made up the KP group. Altogether, there was an impressive bunch of properties.

In 1946, Lawrence Kaip and Andrew Ritz staked the best show ever, the MINT claims group at Courageous Lake, and I arranged for them to be transferred to Payne Yellowknife Limited. It took me one full winter to plan for a $46,000 development program. I got my brother-in-law, W.E. 'Bill' Peters and his wife, Florence, to move into our house on Jolliffe Island and run the operation for me. He flew in food and supplies along with a complete diamond drilling outfit, five igloo-type canvas huts and a small caterpillar tractor to bulldoze an airstrip when the surface thawed out.

We found a quantity of gold in twenty places over the two-mile group of claims, mostly in a rusty gossan zone next to a long depression. I believed it to be another fault zone, with shears, just like the Rycon. Some of the samples assayed at 20.28 ounces of gold per ton, while others averaged 11.28 and 13.86 across six-foot widths. These were phenomenal grades. This was ten times as good as what others were mining in the area.

I hired a geologist and an engineer to map the 3000-acre group. We worked like hell trenching and drilling on that property but we lost in the throw of the dice. None of the lenticular veins had a length or depth of more than fifty feet, and no vein had continuous values over a length of five feet. Where values were received, free gold was always present and assays usually ran to several ounces. The gold was lodged in irregular pockets, and the showings were far below economical proportions. I felt crushed. It was a terrific disappointment.

ALICE PAYNE The shareholders expected Dad to find another mine, or at least make a profit selling the claims, and he felt that he had let them down. But he refused to promote the stock dishonestly, resulting in a battle and the subsequent dismissal of one of his first directors, Viola MacMillan. She had been the Secretary Treasurer for a while, but sometime between September and November 1946, something went drastically wrong.

Since Payne Yellowknife didn't pan out, it's sad to think that they raised over a million dollars and it came to nothing. The only winner in the race turned out to be the University of Alberta's geology department, to whom Dad donated most of the high-grade gold samples from his prospecting pits on Courageous Lake.

On the bright side, there was nothing disappointing about Dad's original find in Yellowknife. The mine was known locally as the Con, unless you read the fine print, and the staff had begun to speculate on finding an extension of the Giant property's orebody, newly discovered across the West Bay fault to the north. This fault zone is one of the most striking features of the landscape. On air photos, it looks as if someone simply drew a line across the rocks.

In the winter of 1939, a mineralized shear zone had been discovered near the southeast corner of the Giant property. Dad and the other prospectors had often traded stories about the gold that could be panned at the mouth of Baker Creek, but no one thought that underneath the drift-covered valley it could be so rich. When A.J. Dadson, a geologist working for Giant's Yellowknife Mines Limited, started diamond drilling in 1944, the samples indicated the largest tonnage of gold ore then known in the Northwest Territories.

Cominco geologists saw no reason why this orebody should not continue on the Con-Rycon side of the fault. The gold occurred in the same setting within the ancient volcanic rocks called greenstones. The rocks had been severely folded and faulted more than once. Could they have been displaced by the West Bay fault? Looking at the map, it seemed logical. The big questions were where and at what depth. Even if deeper orebodies existed, would it be economically viable to mine the gold?

The answers meant everything if the mine was to have a long-term future. Neil Campbell, who'd once worked for Jolliffe, was assigned the task of making a map and doing the calculations on the West Bay fault. He started with Jolliffe's original maps of Yellowknife Bay and Prosperous Lake, and did a painstaking job of unravelling the geological history by mapping the surface. He used several diabase dykes as structural markers to determine the amount of vertical and horizontal movement along the fault zone. The conclusion showed the promise of gold at 16,140 feet south and 1570 feet down on Cominco's side.

The difficulties of making this assumption were enormous: five-degree errors arising from incorrect dip estimates could lead to differences of hundreds of feet in the vertical and horizontal displacement. Ultimate proof was found in 1946 when a 2092-foot deep hole was drilled from the 1250-foot level of the Negus Mine, adjacent to the Con-Rycon property, and intersected gold-bearing quartz after 1173 feet of drilling. The long shot had paid off. Second and third holes confirmed the presence of this major new zone, afterwards known as the Campbell Shear.

This brilliant work was a tribute to Neil Campbell's skill as a geologist and Cominco's persistence in mining. It meant a new lease on life for the Con-Rycon Mine, justifying the risk taken by Cominco in 1937, when they committed a mine of unknown potential to production with only a scant supply of ore in sight. Dad realized the implications at once: the Campbell zone dipped west, and any extension at depth would pass through his four Rycon claims.

TOM PAYNE I had been keeping my eye on Cominco since we still owned forty percent of the shares. They had found a big silicified shear zone under a little lake in the number two claim. They drifted into it for 140 feet, and then drilled some lateral holes to see what was there. All the quartz-rich sections ran up to one ounce per ton gold, and even the low-grade schist contained 0.2. There was another mine right there, under the same spot where I'd been romping around! The veins and the shear system looked continuous, which would give it tremendous tonnage.

By 1942, Cominco was milling 350 tons per day of ore that assayed two ounces per ton. It didn't last long. Milling shut down for three years during the war because of a labour shortage. All the able-bodied men had gone to fight overseas. But Cominco kept steady progress with its geological work. They bought the adjacent claims, and were willing to buy or option any other property with promise. I figured they would eventually own the whole place, lock, stock, and barrel. And what a place! The country was hellishly rich.

Cominco were masters at understatement when it suited them. I often got my information from the mining office, and helped things along with a bottle or two. One man told me that Cominco would give their eyeteeth to acquire one hundred percent of Rycon Mines Limited, the joint operating company. There were sections that were so rich they could be used to up the average grade for leaner zones. Then when the sweet spots ran out, Cominco acted as if the gold was gone forever. In fact, it continued about ten feet off the drift and would intersect again. There were a lot of holes in the wall to look at, where the best ore had been extracted, but we would never know it. I thought we should have had our own man down there all the time, someone financially interested in our company.

The facilities, the ore, and the tonnage were all there, but I had to find out everything else on my own. That didn't bother me too much because I knew we had one of the richest gold mines in Canada. It was only a matter of time before they would catch on and exploit the Rycon. My friend Charles McCrea, the President of Negus Mines Limited, sent me a private and confidential letter on October 14, 1948, telling me the drilling results on the Campbell Shear.

CHARLES MCCREA I was advised from Yellowknife that on their 2300-foot level, Cominco intersected high-grade in the Campbell zone about 1000 feet north of the Negus boundary. This would, of course, carry through Rycon and there would come a time when great quantities of gold would be taken off the property.

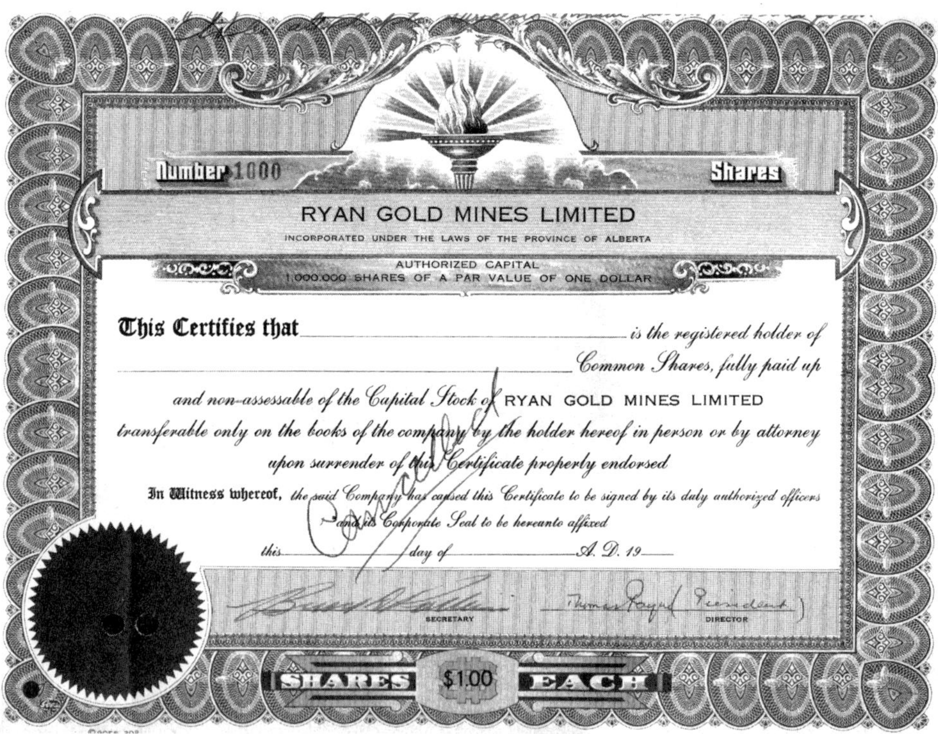

Number 1000 Shares

RYAN GOLD MINES LIMITED

INCORPORATED UNDER THE LAWS OF THE PROVINCE OF ALBERTA

AUTHORIZED CAPITAL
1,000,000 SHARES OF A PAR VALUE OF ONE DOLLAR

This Certifies that ______ is the registered holder of ______ Common Shares, fully paid up and non-assessable of the Capital Stock of RYAN GOLD MINES LIMITED transferable only on the books of the company by the holder hereof in person or by attorney upon surrender of this Certificate properly endorsed

In Witness whereof, the said Company has caused this Certificate to be signed by its duly authorized officers and its Corporate Seal to be hereunto affixed

this ______ day of ______ A. D. 19___

SECRETARY DIRECTOR

SHARES $1.00 EACH

*Ryan Gold Mines shareholder's certificate.*

TOM PAYNE This confirmed all my optimism, and as time went by, the mine looked better and better. The only trouble was that everyone thought I should sell out completely, even McCrea. But the Rycon claims were the foundation of the whole business. Too bad Mickey didn't realize that Bill Jewitt and those company men were trying to pull the wool over his eyes.

CHARLES MCCREA At the time, Cominco was beginning to realize that they had a great potential mine. Their objective was to acquire the interest of Tom and the Ryans, so they could have a free hand. Tom was in an excellent bargaining position, and I knew him to be a square shooter. If it had been me, at his age and health, I would have been inclined to take any reasonable offer. A good cash asset in the bank was not to be sneezed at.

TOM PAYNE I agreed with much of what McCrea had to say, but I thought it should be the other way around, that maybe we should buy out Cominco! So in 1949, Mickey Ryan wrote one scorcher of a letter to Bill Jewitt, then Cominco's manager of mines, to let them know we weren't happy with the state of affairs.

MICKEY RYAN Tom showed me McCrea's letters and underlined all the bits that sounded good. He completely ignored any advice to sell out. So I wrote to Bill Jewitt myself. It'd been more than ten years since they'd taken over our operation and the results were very disappointing. Since Rycon owed more money than when we started mining, our shareholders were confused, particularly as the adjoining properties were reputed to be so profitable. It seemed to me that almost invariably, when a rich vein from either Con or Negus reached Rycon, it suddenly stopped or became very lean.

I let Bill know that we'd received several inquiries from well-known mining companies who wanted our forty percent interest in Rycon, but I always said that we weren't interested in selling. Then I offered to buy out Cominco! Since their mine lost money so consistently, I figured they might be willing to consider selling us their sixty percent. So I said, "If you would put a reasonable price on your shares, we would be interested in buying. Perhaps we will have better luck than you've had, or we might be able to cut expenses enough to make a profit. Many of our shareholders have felt that expenses have been, and still are, too high."

BILL JEWITT I was rather surprised at the general tone of Mickey's letter and could only conclude that either he'd been misinformed, or he'd misinterpreted the news. I reminded him that the original Ryan company was paid $500,000, which was straight profit. We had undertaken an energetic program of underground development, subsequently mining on a limited scale. The 1943 shutdown had resulted in the net advance of $709,050 to Rycon. This included the $500,000 payment, so that the Rycon was developed for an extraordinarily low sum for a property of its size.

A modest operating profit was predicted for the future, or until the higher-grade ore could be extracted. Our operating expenses were high, but the reason was not poor management. Rather, it was the narrow veins plus the usual difficulties of working in the far north. We were not at all ashamed of our record. I'd done a comparison of mining costs between Rycon and Negus and it appeared to me that Rycon's were lower. Besides, Mickey's shareholders had received their money up front. There were occasions when profitable stopes petered out at or near the Rycon boundary, however, this was the luck of mining. So I issued a challenge to Mickey to send someone in to check up.

TOM PAYNE Of course I knew that they had already confirmed the extension of the Campbell shear zone and extended the ore to the Rycon claims. Then about ten months after McCrea's letter, I got a telegram on August 12 from Cominco's head office at Trail and the truth finally came out.

MICKEY RYAN I always felt that Cominco would never mine the Rycon until they owned it all, and that they were stalling, trying to starve us out. None of us were getting any younger. In the original agreement Cominco had promised to "Bring the property into production within a reasonable time, and carry out all its mining operations in a manner which would ensure the maximum extraction of ore from the property, consistent with good mining practice and economical operation." One of the reasons we'd made the deal with them was because they already had their facilities in place. But we had to repay their half million, plus the cost of mining expenditures, before we got any dividends. It was this debt, while little Rycon ore was being milled, that kept Ryan Gold Limited in the red for so long.

TOM PAYNE In 1955, the company took Bill Jewitt's advice and hired Conwest's consulting engineer, R.L. Segsworth, to review the situation. He felt that it would be at least two years before stoping would start on the Rycon into the rich Campbell zone. He pointed out that Rycon's debts to Cominco would increase if Rycon ore was milled

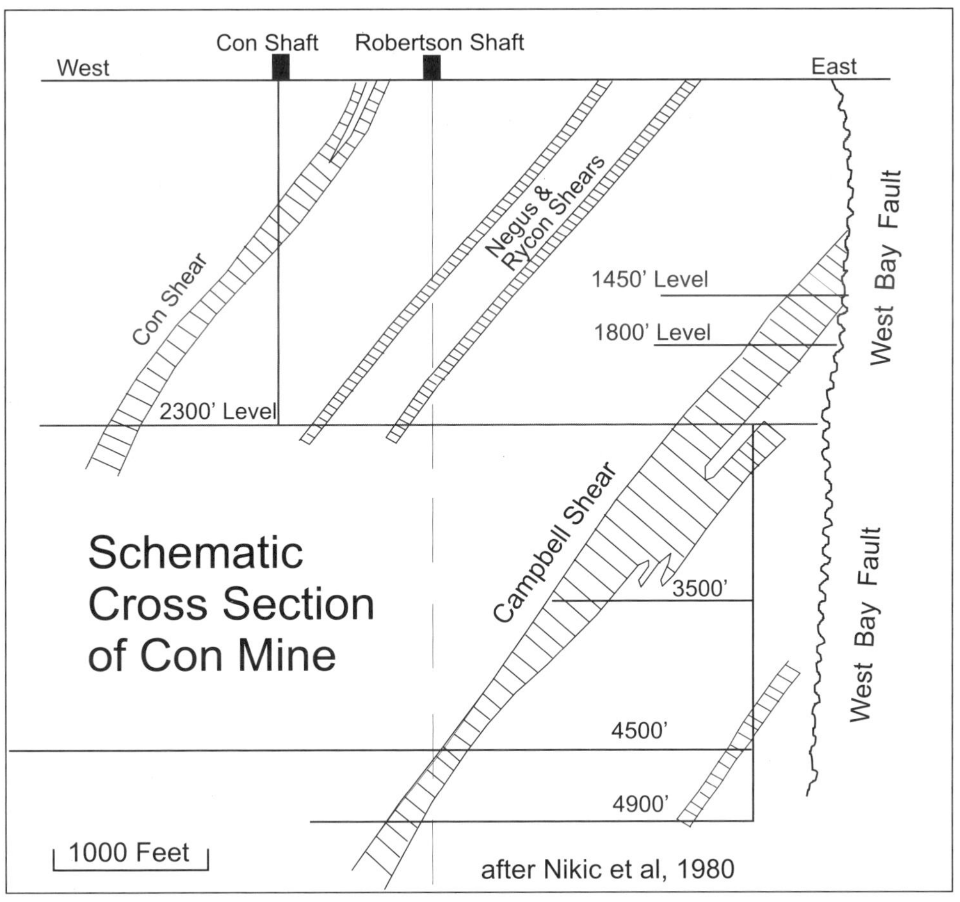

at a rate of only fifty tons a day, while heavy development expenditures continued elsewhere. In conclusion, he felt that under the circumstances, it would be difficult to visualize any profits for the shareholders for several years.

MICKEY RYAN That confirmed it for me. The only thing that stuck in my craw was that Archibald guaranteed that we would get the ore milled on a fifty-fifty basis. We shook hands on the night the original deal was made, which meant more to me than what they wrote down. I wanted Tom to get a confirmation letter from Bill Jewitt that they would guarantee to mill at least two hundred tons a day as long as they had ore reserves.

TOM PAYNE Segsworth really let me down. I didn't agree with his estimate of ore reserves. 112,000 tons was far too low. I believed that mining shouldn't be delayed much longer. My own estimate of ore reserves was 289,000 tons of 0.70 ounces per ton ore with good prospects of mining an additional 420,000 tons of the same grade.

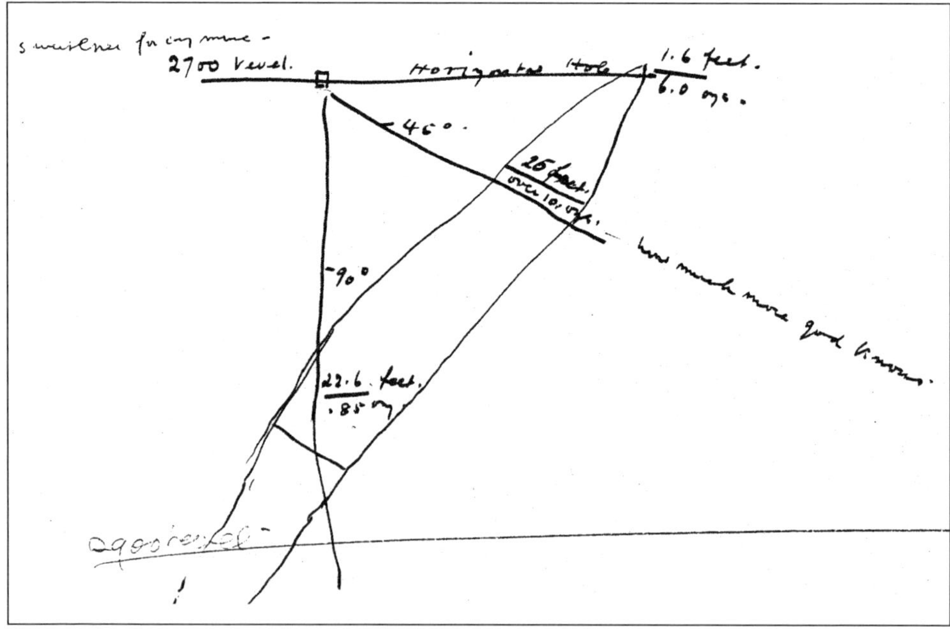


*Tom Payne's rough sketch of the shear zone in a letter to Mickey Ryan, 1957.*

So I tried again, and hired Gordon Brown, the mining geologist at Giant, to keep an eye on the Rycon, and to send me his independent reports. By December 1956, Brown said that there were 351,700 tons of 0.64 ounces per ton in the Campbell Shear, and that one particular section containing the highest grade ore in the Campbell zone would continue at depth into the P&G No. 3 claim. Mining it was inevitable.

I couldn't persuade Mickey to keep his stock any longer. He sold out to Cominco without even consulting me. He just needed the money. After he read the line, "Much too high a value is being placed on the possibility of getting more ore," he bailed out. On top of Segsworth's report, Bill Jewitt had written what I always called the thirty-five cent letter. He arrived at a figure of only 35¢ a share for Ryan Gold. Later I understood how he'd managed to jump up to $1.50 for the official offer in no time flat. My own estimate of the value of our holdings was $3,750,000. With one million shares outstanding that equated to $3.75 per share.

Conwest already had Billy Wilson's shares, which left me as the largest single original shareholder. I kept the faith. They couldn't offer me enough money, no matter what happened. I knew the facts, and it was up to me to keep people honest. I always liked and respected Bill Jewitt. He was a Cominco man, but he couldn't help that! Bill never could understand how I managed to ferret out my information, especially about good ten ounce drill holes. And any time he asked, I'd just say, "Beer Parlour!"

Once on the way to Yellowknife, I went up front to sit with the pilot. He had the radio turned on just for fun when suddenly a coded message came through on a private frequency. He'd been a top code cracker during the war, and could read almost anything as easily as you could read your Corn Flakes box. He took a piece of paper and wrote it all down, then handed me the translation chuckling, "You'll be interested in this one!" Was I ever! It was from Trail and it instructed the staff in Yellowknife to show me around and let me look at anything I liked, except the contents of File 2 in Drawer 3.

Well, when I arrived, I stayed in the guesthouse and had a good squint around. When they asked me if there was anything else I wanted to know, I said I would like to look in File 2, Drawer 3. Of course, they had to show me and risk the wrath of God, but they sure couldn't figure that one out! I asked for a sample of the ore they had drilled into; beautiful stuff, more than ten ounces per ton gold, over a twenty-five foot width.

All this was part of the game. I raised hell with Cominco, and they deserved it, but after all was said and done, we'd made a great deal. They brought the mine into production and kept it going when others might have folded. For my part, I always tried to be honourable and decent in my dealings, something that had been drummed into me at boarding school. I tried to carry out my obligations to the Ryan Gold shareholders, and it turned out all right in the end.

ALICE PAYNE 1958 was a critical year for Ryan Gold because the deal was up for renewal. No dividends had been forthcoming, and Dad was afraid that the shareholders would sell out, allowing Cominco to clean up. He'd sent them all a letter explaining that Ryan Gold was in a profitable position at last, and that there was enough ore ahead of the mill to maintain production for six years at a rate of 200 tons per day. More ore would materialize as the work progressed, and the mine was about ten percent developed and explored. But he knew there would be a showdown with Cominco, and he needed the support of his loyal investors.

TOM PAYNE I told them that there was going to be a snap meeting in April. It would be the first in more than seven years, and I would be sending them a proxy. If they couldn't attend, I asked them to sign over their vote so that I could represent them. I reminded them that as the largest shareholder, I owned 130,000 shares, and that I had still had faith in the property. I promised to look after their interests as I did my own, and pleaded with them not to forget to sign the proxy and return it promptly. There were many other larger interests that could have bought us out, and gained control of Ryan Gold, if we hadn't shown a united front.

ALICE PAYNE Dad had added up all the votes he could count on. Whether by trust, or the romance of holding shares in a Canadian gold mine, many of his English relatives and friends still believed in him; even Leonie, who eventually married an engineer. Conwest, Pat Ryan, and the legal partners were not as secure. He was worried that any one of them might defect, which would leave him a few thousand votes short. But he had a secret weapon up his sleeve: the Catholic Church! In 1939, about the same time as the Rycon deal, Dad had donated an entire operating room, equipped with an x-ray machine, to the Church hospitals at Fort McMurray and Fort Providence, and they never forgot his generosity.

'TOMMY' PAYNE  I remember the meeting because it was the only one I'd ever attended before being packed off to Ridley College, a boarding school in St. Catherine's, Ontario. I was only nine years old and Dad parked me in a corner to watch and learn. We started out at the Friedman, Lieberman, and Newson law office, and then joined the others. Soon they all got busy counting proxies. It was about nine in the morning, and the meeting was due to start at eleven. It turned out that they were about 3000 votes short. Dad had already written to

*Tom Payne and son 'Tommy,' 1958. Payne family collection.*

the Oblate Fathers, and they had promised to send someone if he needed help. So he went over to the seminary, and returned with one of the priests who took a chair out in the hall. Dad said he would give him a sign when to come in.

Meanwhile, the company brass arrived, including a bunch of top guns from Montreal, and a few extra pen pushers from the CPR, which had controlling interest in Cominco. Everyone took their places around the table, and laid out all their papers. They told Dad they were planning to elect their own slate of directors, based on the numbers of proxies already filed. They believed it to be a done deal. Dad just looked around and said, "We'll see."

When it was time for the official portion of the meeting, Dad said that he wanted to check if any more shareholders had arrived. He left for a moment and nodded to the priest, who strolled in and announced that he represented 3000 shares to be cast in favour of Tom Payne.

I can still see the looks on the faces of the Cominco men. One of them jumped up from the table, then they started folding up their papers, and shuffling them together. Boy, did they look mad! Dad just sat there and declared the proceedings open for business. I was just a little kid, so I didn't realize all the ramifications, but it sure made a big impression on me. I can still smell the smoke of Moe Lieberman's cigar.

TOM PAYNE After the battle with Cominco, I wrote a letter of thanks to my faithful shareholders. At long last, I'd finally run Cominco into a corner, and the mine in Yellowknife was now being fully worked. It didn't have the biggest tonnage yet, but the property was only ten percent developed. Without a doubt we had in Ryan Gold the richest gold mine in Canada. At the time, I figured it was worth almost 14 million.

ALICE PAYNE As a group, the Ryan Gold shareholders threatened to cancel the agreement with Cominco when it came up for renewal unless they agreed to mill ore from the mine at a higher rate. Under pressure, they agreed. By mid-year, profits from the mining operation were sufficient to liquidate Rycon's entire debt to Cominco. A reserve cash surplus was built up, which enabled all the shareholders to receive their first dividends in 1959. Six years later, $3.27 per share was paid out, almost exactly the amount Dad had calculated. Within ten years, the ore reserves would stand at a record level at the Con-Rycon Mine. Milling would increase, and a new shaft, the Robertson, would reach the workings a mile underground.

# Black Gold

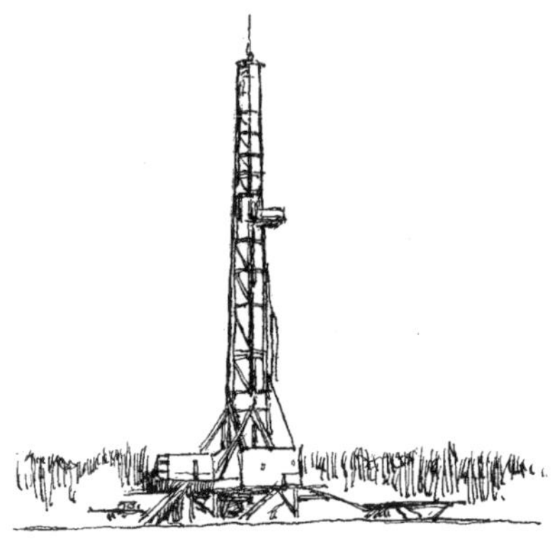

## Chapter 11

Alice Payne The realization that Payne Yellowknife Limited would not produce another great mine occurred the same year that the Leduc oil discovery happened seventeen miles south of Edmonton. When Imperial Oil brought in their well in 1947, Dad was there, and he made up his mind to go into the oil business too. Now, he had to learn the technicalities of oil leases instead of mineral claims.

If you compare a recent map of Alberta to one from that era, it's hard to imagine how someone would have had the courage to proceed. Dad's map, on a scale of eight miles to the inch, covered a good section of wall space. But the known oil and gas wells looked like isolated measle spots on an empty background. In the east, near Lloydminster, and to the south between Lethbridge and Medicine Hat, the spots were slightly closer. A few more wells were indicated in the foothills west of Calgary. The northern half of the province was mostly blank space, except for a small cluster at Fort McMurray, a few near Peace River, and maybe three dozen more scattered north of Athabasca and Lac La Biche.

Most of the oil and gas discoveries on the prairies were concentrated in the younger Cretaceous sands overlying the buried Devonian reefs. Leduc changed all that, and when the Redwater field was discovered in the summer of 1948, Dad drew a line on his map stretching from Redwater to Leduc, and south toward Stettler and Big

Valley, and called it his hunting ground. He put over 160,000 miles on his car looking for suitable land to drill on.

TOM PAYNE There were millions of potential acres stretching from the Northwest Territories to the United States, with more specific sites that were considered hot spots. The competition for these lands was terrific. The large oil companies hired a lot of men who spent their time just correlating the wells, or doing the seismic and stratigraphic work to figure out the geology. Small operators had a much harder time. Although there seemed to be unlimited money for oil exploration, finding good ground was difficult.

Sometimes you could pick up reasonable leases at Government land sales, but in order to do that, you had to produce the cash. Then if the government thought your bid was high enough, you won. If not, they said, "Bid Rejected," and refunded your money. Several times I had a chance at making a private deal with the landowner after obtaining some good information on a well or a locality. But by the time I made financial arrangements, the news had leaked out via the grapevine, and the owner either backed away, or accepted someone else's offer.

Another way to get a lease was through a farm-out from one of the bigger companies. When a company drills a producing well, it's often forced by law to drill on the adjoining forty acres. These are called offset obligation wells, the idea being to prevent other operations from siphoning all the oil out from under without paying the real owner. Of course, if they think there is nothing there, or the risk is too high, or they run out of time or money, the company will let someone else take the chances.

I contacted Imperial Oil as soon as their head farm-out man showed up, because they were giving away some very attractive ground near Leduc. In the end, I decided to get some freehold land from a farmer. I turned myself into a first class lease-hound, and drove all day and all night looking for chunks of moose pasture to buy for a few dollars. At first I bought anything I could for thirty-five cents an acre, and turned down all sorts at fifty cents, but after a while, two dollars an acre began to look cheap.

My only trouble was that I had a hell of a time talking to some of the farmers. Once I went to buy some royalties from an old mossback down at Leduc, but he told me, "You're too late, I'm in partnership with Imperial Oil." Imagine that, him thinking that was the case. Another time, I sent John Larsen, a driller I knew from Yellowknife, out to get a quarter section of land near Leduc. An old Russian owned it, and there were rigs all around his land, but he wouldn't sell. He had a cow, a calf, and four horses walking around, and lived in a ten by twenty foot house with a sod barn out back. All he ever seemed to eat was porridge.

JOHN LARSEN Tom rigged me up with an old car and some gear, and sent me down to try to make a deal with the guy. I had nothing else to do, and Tom told me not to rush things. I bought some groceries, got an old Chev down at the garage, which Tom paid for, and I drove out there. "I'm broke," I said, "But I have my grub and stuff, and I wonder if I could stay with you overnight?" He said, "Yeah, okay, sure."

Anyway, the farmer was a religious son of a gun. He had a bible lying on the table, open and upside down, and when I saw it, I knew what subjects to avoid. And I sure couldn't talk about the oil rigs until he talked about them. He cursed them, and I agreed with him. There was nothing else to do. So it came down to this. I had to make an end of it, and I said, "By God, you own the mineral rights on this quarter section, why don't you sell them? Maybe you've got no oil, but you can find that out, and afterwards, you'll get a royalty or you could sell this place, maybe for ten thousand dollars."

He looked at me and laughed. "Jesus Christ, you! A tramp? Talking about ten thousand dollars! Where would you get that kind of money?" I kept trying to convince him, but nothing doing, and finally he chased me away. Had I gone down there in Tom's Cadillac, all dressed up, the same guy would have still run me off with his shotgun and called me a crook. You just couldn't talk to him. He'd made up his mind that if God had put the oil there, he meant to give it back to God. The trouble was that he was surrounded by all those Imperial Oil wells that were sucking him dry!

ROBERT FOLINSBEE Many of the freehold farmers were reluctant to lease their land. They'd worked it for fifty years since the first wave of homesteading. They knew the value of labour that had been blandished on it, and were cautious about signing over their oil and gas rights to smooth talking landmen. Tom was a more interesting character than most. He'd suffered through tough times too, and they liked his blunt good humour and down to earth mannerisms.

TOM PAYNE At last I got my hooks into a real good oil show within a mile of Imperial's discovery well at Excelsior. I was lucky. By the end of 1948, companies had only drilled eight wells in the surrounding 900 square miles, including an old historical test hole drilled in 1913. A year later, there were eighteen more, and things were heating up. This time I sent Alan Snell, my stockbroker at the firm of Bongard Leslie, to sign the long awaited lease with Mulligan, the farmer and landowner.

INCORPORATED UNDER THE LAWS OF
THE PROVINCE OF ALBERTA
AUTHORIZED CAPITAL
4,000,000 SHARES OF NO PAR VALUE

No 484 Shares 1

METRO OIL & GAS CO. LTD.

THIS CERTIFIES THAT ------BRUCE D. PATTERSON, Executor of the Estate of Thomas Payne, deceased------ is the owner of Shares of the Capital Stock of

METRO OIL & GAS CO. LTD.

transferable only on the books of the Corporation by the holder hereof in person or by Attorney upon surrender of this Certificate properly endorsed.

In Witness Whereof, the said Corporation has caused this Certificate to be signed by its duly authorized officers and to be sealed with the Seal of the Corporation this 16th day of October A.D. 1968

PRESIDENT SEC. TREAS.

SEAL

SHARES NO PAR VALUE EACH

COUNTERSIGNED: FRIEDMAN, LE[illegible]MAN & [illegible]ON, TRANSFER AGENTS, EDMONTON, ALTA.

BY

*Metro Oil & Gas Limited shareholder's certificate.*

## Central Alberta Location Map

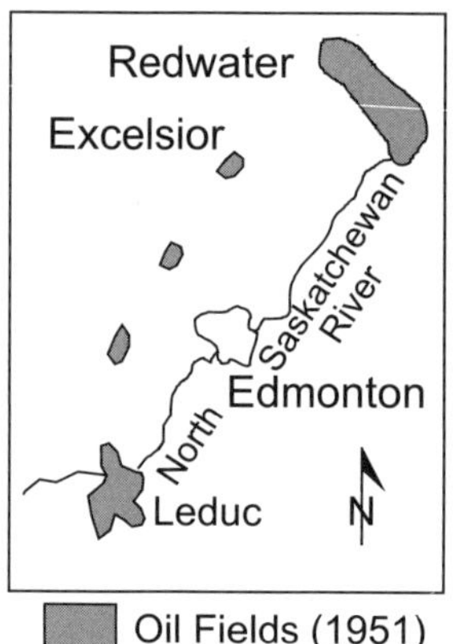


## Excelsior Field

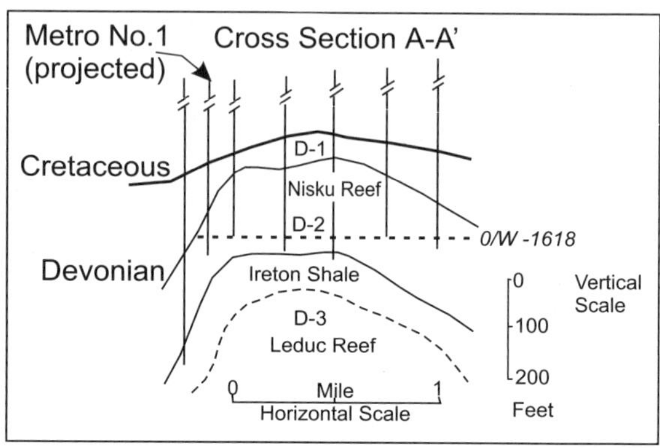


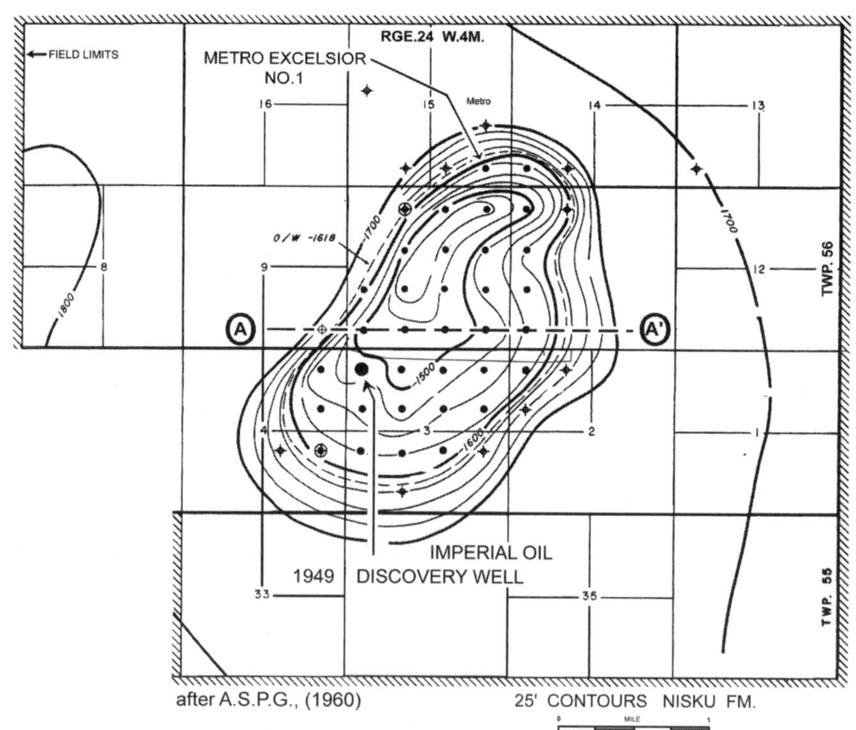

Alan Snell Tom had been scouting around there and was real nervous. He wanted the deal very badly, and he was afraid that Mulligan would turn him down. So he came to me and said, "When I talk to farmers, I get upset. You're more patient; you go and talk to him. There's a real good prospect, a good lease, out there." So I spoke with Mulligan for quite a while, and he finally agreed to sign up with us.

I had to go back to town for the legal papers, and when I returned, Mulligan met me at the door: "Since you left, there's been about five oil scouts come out here." And I thought, "Oh no, there goes our deal." But then Mulligan assured me that we were still on! So we sat down at a rough table in his wooden shack, signed the papers, and that was it.

He really stuck to his word, the old boy. I gave him a lot of credit for that. Imperial Oil, Shell, Gulf, and their respective scouts had offered him more money than we did. We paid fifty cents an acre for three quarter sections, 480 acres, just for the option to drill. He made the deal with us because we were a small operation and he thought we were more likely to drill on it, and not leave it sitting there like a big company might.

Tom Payne After that, I took a pair of Zeiss binoculars and spent a lot of time scouting Imperial Oil's wells, to see when, where, or if they struck oil. I'd count the length of pipe stacked on the derrick to determine the depth of drilling. I'd also watch for the drill stem tester's truck to arrive. If I was lucky, I could see oil flowing during their test.

It was a sneaky business. Often the oil companies wrapped their rigs in canvas, or they'd pull pipe out of the hole to fool you about the depth! Sometimes, before a land sale, they would run casing on a dry hole and pretend to cement using drilling mud, acting as if it were a real well. It was some game, a real science! All the roads in and out of a well site were blocked off. No one could go in or out, and if they could, they'd confiscate the crew's car keys. No one could drink beer or do a darn thing. And if they caught you hiding out behind a willow bush, they would throw you in the sump.

Thank God that never happened to me! Later, scouts would fly around in helicopters taking photographs and looking at the weight indicator through a telescope. They can figure out the depth to the nearest five feet! It costs a lot more but you can't be thrown in the pit if you're up in the air. I didn't have to go very far for my scouting job at Excelsior. Imperial spudded two wells across the road to the south of our lease.

ALAN SNELL When the Imperial Oil well came in, at five-thirty in the morning, Tom was out there for the test. He knew about it somehow. Then, at about nine o'clock he walked into my Edmonton office at Bongard's, jokingly called the bucket shop. With a gin bottle full of oil, he said, "Well. That's it."

TOM PAYNE Just before we drilled, I ran into the Imperial Oil geologist in the beer parlour, and asked him what he thought. They'd drilled on all four sides of our lease, but two of their wells were no good. The other two were key wells, but I didn't know which one was the best. We had a choice of twelve possible sites. He pointed out two likely places on my map. So I made up my mind to drill as far away from those locations as we could get, thank God.

None of the eastern mining men would put up the money to drill a well, so we had to raise it ourselves. Snell and I formed a syndicate called Metro Oil and Gas Limited, along with Melvin Friedman, a lawyer, and Harry Cohen, the owner of the Army and Navy stores. Then we hired Commonwealth Drilling to bore a hole for us at five dollars a foot. They moved in with a crew of roughnecks and a toolpush, spudded in on April 28, 1950, and worked around the clock. The roughnecks got seventeen dollars a day, and the toolpush, or foreman, got twenty. We hired a geologist to look at the rock chips and cores. I called him the Pelican.

I had the tops of all the geological formations, and I examined all the samples when they came out of the hole. The geologist would inspect a little bag of rock chips from the drillers, and wash them off, and take a look under the microscope. I kept notes on everything. At the top of the Viking, there were little chert pebbles. Then there was the Glauconitic sand, which was greenish; it looked like copper stain to me. Next were the Ostracods, which looked like little beetles with no heads. The last formation before we reached the Devonian was the

Basal Sand, with quartz crystals cemented together. Finally we hit the residual zone at the unconformity, and the top of the first Devonian carbonate, the D-1.

We had drilled through a silt zone, and some red beds, before I got really worried. In the Imperial well, they didn't go through such a thick section to reach the second carbonate layer of the Devonian, or the D-2, which was what we were searching for. I had a little tent, and pitched it right there at the well site, just to keep an eye on things. One morning, at six, they drilled into a zone of limestone that smelled like petroleum, and I raced to Edmonton to find our geologist.

An RCMP stopped me for speeding. I must have been doing ninety miles an hour, but he wasn't worried about that. He looked at my shirt with the elbow out, and my beat up, dirty, prospecting clothes and

*Drilling rig at Metro-Excelsior #1, with Tom Payne's tent pitched on the lease. Payne family collection.*

asked me if I owned the car, my 1947 Cadillac! "Yes," I said, "And what's more, it's paid for," and I produced the registration. The next question was, "What nationality are you?" So I told him I was English, but I couldn't help it. So he let me go.

It took three weeks and $30,000 to drill that oil well. The last twenty-four hours before the well came in, nobody slept. That was some excitement! Once we were in the D-2, we cut some core. It had wonderful porosity and the oil just seeped out. We ran the casing and pumped down liquid cement between the pipe and the side of the drill hole. It set solid within a day. Then we shot the pipe and cement full of holes with a perforating gun so that the oil from the formation could run in. We poured in a couple of thousand gallons of someone else's crude to control the well.

The well blew in on Sunday morning, May 25, 1950, at about three or four o'clock. I had bottles of Mumm's champagne tied all over the willow bushes for my friends. Dark brown liquid shot up more than

*Oil well blowing in, May 25, 1950. Payne family collection.*

two hundred feet in the air, spraying everything for hundreds of yards, even Mulligan's house. I waded right in. It was beautiful and sweet smelling, just like good crude oil should be. I got oil all over my clothes that day, but wouldn't let Olga send them to the cleaners.

ROBERT FOLINSBEE It was risky business, and Tom had pondered his chances. He had narrowed it down to four potential locations: two were close to Imperial's discovery, and two would likely go down the slippery path to ruin. I've seen the deep disappointment of small entrepreneurs like him, who had lease holdings on the edge of major reefs. They would watch the drill bit go down to, or below, the oil-water interface. Then they would see the geologist anxiously peering at samples from the shale shaker and observing only the capping green shale, not the porous oil-stained dolomite of the producing reefs.

Talks with Imperial geologists suggested that one of Tom's locations would be better than the other. But Tom's distrust of big companies led him to chose the least favourable. He babysat the well as drilling proceeded, and held his breath while the drill approached critical depth. The right chips appeared thirty-five feet above the anticipated oil-water level and a drill stem test proved a good commercial well, blowing smoke rings from the flare pit.

TOM PAYNE We toasted our success, and I got Mulligan out of bed. He answered the door saying, "I suppose you've come to buy me out!" I told him, "No, I've come to tell you that you're a rich man because we have got an oil well! Don't sell any of your twelve and half percent royalty rights whatever you do! Here's a cheque for one thousand dollars so you can eat. It's not a gift, and you can pay me back later, because you should get a good price for the oil you have left."

HAROLD KESNER It was a pretty big thing for the Mulligans because they didn't have a dime. Their place was nearly beyond description. None of the rooms had doors, there were dirt floors, and the chickens, pigs, and dogs walked in and out from the barnyard. The house had no paint, and no furniture, just a big brass bed with Home

Sweet Home tacked on the wall above it. We sat on boxes whenever we came to visit. No one believed me until one day I took Paul Carpenter, another stockbroker, out there for tea, which was served to him in a tin can. Some of the Mulligan's first oil money must have gone to buy dishes.

FRANK CLUNE I met Oil-King Tom on my travels through Alberta. He was six feet tall, fifty-eight years old, and wore a red checkered shirt and glasses. He struck me as a man who had done many things before going into the oil business. Together, we drove through a district of wheat farms during planting season, and stopped at Mulligan's farm. Their little house, sheltered by poplars, stood just a hundred yards from the new Metro-Excelsior Number One well. The new gusher had been sealed, pending the erection of storage tanks.

*"Sudden oil riches are taken in stride by Mulligan family of Bon Accord," by Eric Young, from local newspaper, 1950.*

As I travelled in Tom's company, I began to understand that gold mining and the oil business are not the result of blind chance, like a win in a lottery. Real wealth is won from the soil by hard work, plus knowledge and skill. So I didn't envy him, because he'd earned every penny.

Tom Payne We planned to drill more wells on Mulligan's land, but before they were spudded in, he had sold most of his royalty rights for about $10,000 each. He bought a new washing machine for his wife, and a Chrysler so the family could holiday in California. He spent the rest on building a new house, and buying a diesel caterpillar, a three-ton truck, and a few things for each of his eleven kids. There was a big article about him in the Edmonton papers, complete with a family picture. He didn't mind the publicity. The only thing he regretted was that he couldn't wash the car for the photo.

Saving money meant little to the Mulligans, and I started thinking that I should collect my one thousand before it was too late. He told his wife to "Write Mr. Payne a cheque," and I took it straight to the bank. When the girl brought out the account sheet, I looked over the top of the wicket and read that out of the $70,000 or more they had received, Mulligan had only $3000 left!

Alice Payne That was a rare good well in the D-2 reef, fifty-five feet of pay, and a steady producer. It turned out to be a separate reservoir from the main Excelsior reef, a small pinnacle, and it pumped oil for about thirty years, producing over 600,000 barrels. Metro drilled two more wells at the Bon Accord site, but they were dry holes, or dusters. He actually did Imperial Oil a favour by defining the northern limits of the Excelsior field. In 1964, they offered to buy Metro for $125,000, but Dad turned them down. When they asked why he hadn't even looked into the deal, he told them, "It's not every day you get to say no to Imperial Oil!" He figured there were still over 200,000 barrels left.

Dad's only other oil venture resulted in a one-day well out near Skaro, northeast of Edmonton. Trying yet again to make a dollar for the Payne Yellowknife shareholders, he acquired a lease and drilled down to the Cooking Lake, hundreds of feet below the D-2. He left the well on a Saturday night after swabbing out 175 barrels of clean oil, but when he returned in the morning, there was 1000 feet of salt and sulphur water from the carbonate platform below. The layer of oil was too thin, and an ocean of salt water lay underneath. That same day, Dad wrote *Quod Erat Finitum* in his journal, said goodbye to the oil business, and stacked his well cores on the mantelpiece.

# Reflections

## Chapter 12

Alice Payne    Dad regarded his hunt for oil in the same spirit that he searched for gold, only the rules differed. No wonder that author Arthur Hailey found Dad irresistible, and used him as a model for his prospector in the novel *Hotel*. He always insisted to census takers, election enumerators, and the passport office that his profession was 'Prospector,' and he urged all his friends to get out there in the field instead of wasting their time on summer holidays. A favourite line was, "You can stake all the mines I left behind."

His adventurous life became legendary. In an article in the Yellowknife newspaper, written by the ultra -conservative Gordon Sinclair, he was touted to have walked over the entire Northwest Territories in a twenty-five year stretch without a dog team. Dad claimed that he'd never been attacked by a wolf pack, or stalked by a polar bear, and survived for seven years without eating a fresh fruit or vegetable. During his early days of trekking over the barren lands, he sometimes went a full year without seeing another human. Being rich hadn't gone to his head either, a quality that the local people admired the most.

Filling in the time gaps of his life was tough. It wasn't that Dad wouldn't talk about things, but that he would talk too much and go off in tangents. Every time he was in full flight, in the middle of his northern stories, he would pause, and then say, "I went everywhere, from the west coast to the east coast, twice. Walked across the barren lands, damn it! Living off the land, making my own pemmican, making everything. Holy God! In seven years, I never saw another white man. It's all on the map in my bedroom." He insisted that because he could hunt seals, fish, shoot, and survive where others had starved to death, Chief Old Man Zoe had made him an honourary member of the Dogrib tribe.

It's likely that he went trapping with the Semmlers at Aklavik along the Arctic coast, and hunted on the ice with the Klengenbergs at Coppermine, but the evidence has disappeared. Nonetheless, he was convinced that certain place names were bestowed in his honour, such as Payne Lake, and Payne Bay on the Ungava Peninsula. Considering his adventure on the ice floe, we can never know for sure. That another lake bears his name, located northeast of Yellowknife, is a more plausible story.

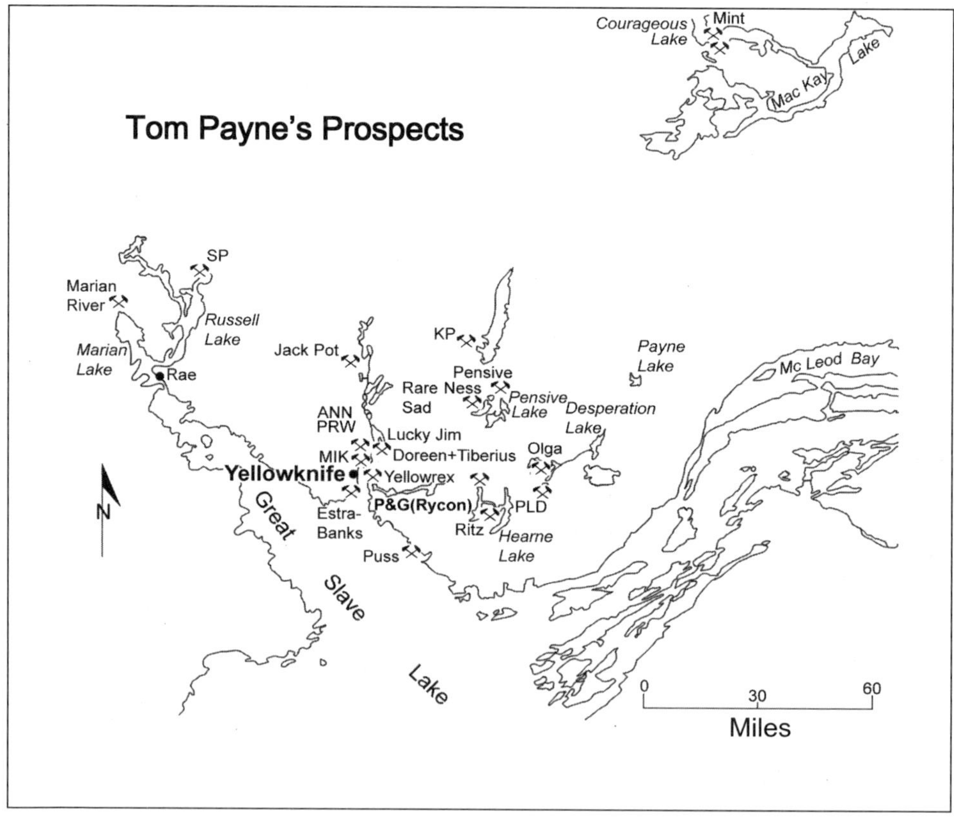

But there was nothing unreal about the impression he made on others with his personality and physical presence. It was overwhelming. His girlfriend in England waited over twenty-five years to marry him. In Saskatchewan, the English relatives and the farmers remembered him walking into their kitchens, fresh from the north. His six-foot two-inch frame, enlarged by his fur suit, took up a lot of space! They thought that in one lifetime, Tom did what five ordinary men could do. The people he met, from Aklavik to Coppermine to the shores of Hudson Bay, thought enough of him to buy shares in his company while the country was enduring the Great Depression. Many of them kept their shares until the company was sold. There must have been some truth in the legend.

Dad's wanderings remind me of the voyageurs and early explorers who covered vast distances and never stayed in one place for very long. But at least they stuck to one profession, unlike Dad, who couldn't even stick to one mineral commodity. He was a restless, driven man. It was more than just wanting to prove himself. He described his life as "brilliant in flashes, but dull in shots" and admitted that he dabbled at one thing after another, rather than applying himself to become a power of good in the world. Perhaps his donation of two operating rooms to the Church hospitals gave him the sense that he could finally give something back to the north.

At the same time, he wanted to show his family that he'd found success in the new world. When he was in his seventies, he travelled to England for one last visit. I had already booked a Youth Hostel excursion with some of my friends. True to form, he pre-empted my trip by coming home with two tickets for the Empress of France, and then ordered a brand new Jaguar XK-150, to be delivered in London. He introduced me to the Pawsey relatives, and I drove Dad around to visit all his old friends.

RUTH MACLENNAN Being a Pawsey, Tom considered the Canadian branch of the family to be relatives. His idea of an evening's entertainment was to talk and drink Dewar's Ne Plus Ultra scotch. At the house in the Highlands, or later when he moved to Saskatchewan Drive, he would produce a bottle and say, "You never go home until it's empty." And he would get out the soda water, a bucket of ice, and four crystal glasses, and the four of us would sit there and talk and demolish that bottle. He poured himself singles and everyone else doubles!

I first met Tom when he got released from hospital. I was single then, and Tom and Olga were very good to me. I had a roommate, and we were living on pennies in a little suite above a hardware store. Tom used to surprise us by bringing us a treat. Usually it would be a big hunk of cheese, or a fruitcake, which were absolute luxuries. He realized that we were hungry, and he felt very fatherly towards us. He used to supply us with ducks; that was his best treat. He'd just waltz in and throw them on the table.

One time Tom phoned to say that he had some ducks for us. My roommate and I were in a hurry to go out for the evening, so Tom said just to leave our door open and he would deliver them. So we left the door unlatched, and we both hopped into the bathtub to save time. All of a sudden, the bathroom door flew open, and there was Tom. He threw those mallards, feathers and all, right in the water with us! We yelled and jumped, having been taken completely by surprise. We still laugh at it, and that was over fifty years ago!

Tom was a practical joker, and once when he saw Havelock Maclennan's car outside my apartment at night, he sneaked up the stairs and hung a sign on the door that said, "Keep Out – Man At Work." When Havelock left for home, he saw the anonymous sign and was pretty embarrassed. He knew he would take an awful ribbing at the Baker clinic from the other doctors, because he was known as the best catch in town. So he studied the note, thought about it, and tried to figure out who wrote it. Finally he phoned me back and said, "It must have been a patient because the name of a drug is written on one side of the page." Of course, I knew right away who was the culprit. Havelock was one of the doctors who had treated Tom at the Royal Alexandra Hospital when he was so ill.

Tom was very observant, and seldom missed a thing. Once, when I was admitted into the University Hospital, he came to visit. I would order lunch from the kitchen, and have it delivered. The nurses thought he was my dad. One day, a pretty blond nurse came in to take my temperature. When she left, Tom said, "She's a Métis." I replied with, "Oh no, she's not. She has a Scandinavian name and her complexion is quite fair." So we made a wager. That night, I asked her to tell me about herself. When she began, "I was born way up north," I knew my bet was lost.

She explained that her father was a white man who had been a trapper and taken an Indian wife. As their daughter, she had also married a trapper, but he had died prematurely. She was taking nursing training to support herself and their ten-year-old son. The next time Tom came back to visit me, I gave him his dollar and asked how he knew. "You just watch her," he said, "When she comes in, she'll do the teepee duck." And sure enough, he was right. She made an almost imperceptible bow that surely came from spending childhood summers in a teepee, and winters in a log cabin with a low door.

Tom's necktie was quite the conversation piece. It had yellow and red stripes on a navy blue background, but was all torn and bedraggled. When I asked him why he didn't get another, he said it was his Felsted school tie, and he couldn't get another. So I made some inquiries at the hospital's occupational therapy department to see if they could weave another. They told me there was a loom at the veteran's unit at the old Government House. So when I was home from the hospital, I took Tom's tie over and left it, so they could get the colours right. Well every day, Tom asked where his tie was. Finally, about ten days later, we got in his Cadillac and drove over. He parked the car while I went to see what was taking so long. The woman I talked to looked quite blank at first, but when I described the tie, she said, "Oh yes, I remember now, one of the patients hooked it into a rug!"

Well, I felt awful. I couldn't go out and tell Tom the truth. When I got home, I phoned my cousin in Montreal, and asked him if he had an old Felsted tie. After he stopped laughing, he said yes, and agreed to send it out by airmail. He had attended Felsted too, and could always get another if he wanted. I didn't tell Tom that the tie came from Percy. I took off the label and Tom remarked so often, "Best damn weavers I ever knew!"

NELL WHEELER Tom used to drop by our house sometimes in the morning. My husband, Ben, was a doctor too, and would leave early for the hospital. So I would often sit and have coffee and a cigarette, and Tom would show up. I was always glad to see him because he was such fun. He said what he thought, and did what he wanted, and what the neighbours must have thought didn't bother him at all.

One day he turned up at our house with his pants ripped and asked me to fix them. I often did little chores for Tom, but I'd lent my sewing machine to my sister Ruth. So I told him to go over to her house. Ruth had employed a housemaid, a well-mannered country girl called Ruby. Tom burst through the front door without even knocking. When he asked for Ruth, Ruby replied, "She's upstairs." "Well," said Tom, "I need someone to mend my pants because I'm not going anywhere like this." Then he bent over, and his pants were split right down the seam. He peeled them off right there in the hall and handed them to Ruby saying, "Go tell Ruth to fix them up."

Well, Ruby was completely and utterly horrified. There stood Tom in his long underwear and she ran upstairs to Ruth and cried, "There's a strange man taking his pants off downstairs and I QUIT!" It wasn't the first time that Ruby had seen Tom, but she didn't know him very well and was a little bit scared of him because he was so big. Ruth calmed Ruby down by saying; "There's only one man who'd do that, and that is Tom Payne." As soon as she saw the size of the pants, she knew she was right. So Ruth did the sewing and lost her maid the same day!

We could never get enough of Tom's northern stories. One morning, Harry, my oldest son, was home. He'd been taking geology courses at the University of British Columbia, and was working at Edson, west of Edmonton. We'd put him and some friends up over the weekend. There was Fred Farkas, who later started up an oil company with Harry, and two girls they were particularly fond of. We'd had a wonderful time Saturday night, but on Sunday morning, Fred and I were the only ones up. Ben was away at the hospital, and we were getting acquainted over a cup of coffee. It was only about nine o'clock, and in walked Tom. So I said, "You know, Fred, Mr. Payne was a prospector."

Fred was instantly captivated. He and Tom started to talk, and I don't think Tom stopped until after one o'clock. Fred was a good listener, and had a knack of asking the right questions. Pretty soon everyone else was drawn in too. That's when we should have had a tape going. Later, Fred said he didn't know when he'd had such an entertaining time, or met such an interesting man. When Tom started talking about mining and oil geology, he knew more than Harry and Fred put together.

It was Tom's personal abilities that made the difference. He was lucky, but he could also recognize something of value when he found it. He sure didn't pussyfoot around. I remember once when he came over after inspecting his apartment blocks. The caretaker thought he was a tramp, just nosing about, and threw him out. Tom was so pleased that he gave the caretaker a twenty-dollar bill for doing a good job!

He was always giving people presents. He used to write me the nicest letters when I first knew him down on the farm at Edgerton, and sent me a lovely pair of Indian slippers, all beaded and trimmed with white fur. Another time, he ran into a stewardess whose father had worked in the livery barn at Dodsland, where Tom called home in his farming days. He gave her 1000 shares of Payne Yellowknife stock, just for a present. At the time, the shares were worth forty cents each.

OLGA PAYNE I always thought that Tom should have been on the stage. He loved creating a scene. Once when I was baking Christmas cake, I sent him downtown to get some citron peel. It was wartime, and peel was hard to find. The only place you could get it was in Eaton's basement. The manager was a good friend of Tom's. So in he went, and bought the peel. But when the clerk started to put it in a paper bag, along with his sales slip, he said, "Never mind the bag," and walked off minus the receipt, and popped the small package into his pocket. When he reached the door, the house detective tapped him on the shoulder and said, "You have something in your pocket that you haven't paid for," and accused him of stealing it!

Well, that made Tom so darn mad! When he tried to explain the situation, the detective just looked at his old clothes and insisted they see the manager. Of course, when Tom appeared, the manager said,

"That's all right. Mr. Payne doesn't bother with receipts or brown paper bags." Tom always said he made a big mistake by not running away. "If only I'd run, and they had chased me or grabbed me, then I could have sued them for defamation of character. There would have been hell to pay, but I didn't think fast enough." By this time, he'd completely forgotten about the cake.

Legal proceedings had always interested him, and he savoured his run-ins with the police or the courts. He had once been involved in a case in the Alberta Supreme Court, when two men insisted they had sent him to stake his four claims at Yellowknife. After some days of testimony, Tom finally gained the witness stand and asked the judge for permission to ask a question. The request was granted. "If these gentlemen sent me out to stake those claims, perhaps they could tell the court where the claims are located," he said. The two men couldn't, and the suit was dismissed. Afterwards, the judge remarked that it was the first time a witness had ever cross-examined him!

Tom couldn't even deal with a speeding ticket without making a special case out of it. One fall he was on the way home from a mining meeting in Trail with one of our friends. On the way to Fort McLeod, an RCMP officer pulled him over and issued a summons, right on the spot. Tom was probably talking too much at the time, and not paying attention to his driving. He made a terrible fuss, said he couldn't wait around because he had urgent business in Edmonton. Finally the policeman agreed to take him to the next town to find the local magistrate. When they found the house, the magistrate's wife said her husband was out duck shooting and he wouldn't be home until after sundown.

So right away, Tom apologized and said he would leave a fifty-dollar cheque in good faith. He promised to pay more money later if need be. No one knew what the proper procedure was, so he wrote out the note, gave her our address, hopped into his car and left. But on the back, where nobody looked, he wrote, "Paid under protest as the Justice of the Peace is out duck shooting." He knew that duck season wasn't open yet, and hoped that the cheque would be presented for payment at the bank, but it was never cashed.

ALICE PAYNE What my mother had to put up with! I remember the time that Dad left home for three days, and took up residence in the Edmonton Club. The reason? The paper had issued a lot of health bulletins about butter versus margarine, and the experts made a case for using the bland substitute. Mother wanted to use margarine, so Dad moved out.

The Edmonton Club was similar to men's clubs in England. It had an elegant entrance-way, and red velvet curtains covered the glass panels in the foyer. If you tried to push the curtains aside and peer in, you were confronted with an opposite set of velvet curtains on the other side! Women weren't allowed in, and according to Dad, if you were in trouble with your wife or the law, you could hide out there for three days with no questions asked. A valet would see to all your needs, and if you chose to be unavailable, he simply said you weren't in, or "not receiving."

Dad was the Club's third oldest member, and well up in the hierarchy. He supplemented his meals at the Macdonald Hotel, and told Mom that he would come home when she got rid of it. The standoff didn't last long, but neither did his affiliation with the institution. When some of the younger members voted to allow women in for Sunday dinner, Dad resigned his membership. He claimed that the new policy ruined the atmosphere of the whole place.

KEN WHEELER In the fall, Tom would take my brother, Alan, and me out duck shooting. Tom would be so engrossed in talking that he would often drive right through a red light, or absent-mindedly stop at a green light while fellow drivers screeched to a halt, and we cringed in our seats. I only got to know him as an older man, and at first I didn't believe some of his antics. But when I started to work on the boats going down the Mackenzie, and said I was from Edmonton, people would ask about him. They would light up at the very mention of his name, then start to talk, and recount many of the same stories I'd already heard.

I always felt that Tom was one of the world's last rebels. Like others who spent time in English boarding schools, he had an independent spirit and didn't give a sweet goddamn what people thought. Once when he was attracted to the Vicar's daughter, the Vicar wasn't too keen on Tom, and told his daughter to keep her distance. The word was also issued to Tom's father, Dr. Payne. Of course, that angered Tom and the following Sunday, he took his father's horse and buggy and drove it all around the church during morning service. The courtyard was paved with cobblestones and he made such a clatter that he disrupted the whole sermon. No wonder he couldn't get along with his dad. But Tom couldn't wash out his English roots, and he never completely lost his accent.

Every time I walk around barefoot, it reminds me of Tom's story about an Indian girl who came out to Edmonton for some medical treatment. He always kept track of his friends from the north, and she got in touch with him. In the course of their conversation, Tom asked what she thought of the white man's world. She talked about the buildings, the people, and all the exciting things she had done, and said that she liked most of it. "But Tom," she ended, "These white man's trails are really hard." She'd been used to walking barefoot, or in moccasins, on the soft muskeg and the paved sidewalks hurt her feet. He knew exactly what she meant.

Tom never lost his sense of wonder about the simplest of things. It was part of his charm. Once when we were out shooting, it had been raining for two or three days. He couldn't avoid going through the mud puddles with his Cadillac, so every now and then, he would stop at a dry spot, take out a cloth and wipe the windshield. I was about sixteen, and knew about all the gadgets as I was really interested in cars. Finally I asked him why he kept stopping all the time when all he had to do was push a little button on the dash. So I reached over and pushed it, and you could hear the sloosh-sloosh, and the water squirted out, and the wipers went click-click. I'll never forget the look on his face. He stared at me and said, "I'll be god-damned. What will they think of next?" So he had a windshield washer, and was tickled pink about it.

'TOMMY PAYNE' Dad was always taking stuff apart with me. At six years old, he gave me a Meccano set for Christmas. After I got really good with my fingers, I set my sights on our Big Ben alarm clock. I opened the latches, spread out all the pieces, and laid them in an organized pattern on some newspapers. When Dad came looking for his clock and saw what I'd done, he said, "I hope you can get it back together without any pieces left over." And I did – it was great!

When I was old enough to go hunting with Dad, we stopped in front of the Eaton's store to pick up some shells. There was a parking spot right in front of the door, with some time left on the meter, but suddenly a little Volkswagen pulled right in behind us. Dad didn't hesitate! He threw the Cadillac into park, reached over and grabbed an old ax handle, which he kept in the back window. He jumped out into the street and stood behind our car. All the people on the sidewalk had stopped to see if there would be a fight, but the other man didn't even get out. He simply put his car in reverse, and backed out the same way he'd come in. Then Dad parked his car in the vacated spot. The only comment he made about the whole thing was, "That god-damned driver!" The poor guy must have been terrified, but I knew Dad was just putting on a show.

*Tom Payne holding court, 1965. Payne family collection*

I learned how to be fearless from Dad. He could mingle with cabbages and kings, go anywhere, and speak to anybody. If I hadn't had a Dad like that, I'd never have had enough courage to pursue my interests in the railway.

We all expected him to live forever. During the sixties, his health failed rapidly due to the rugged life he'd led. The doctors ordered him to stop smoking, but after a few months of abstinence, he continued his old habit. Forfeiting apple pie and ice cream was punishment enough! I was in a cathedral in Buffalo, New York, on October 16, 1966, with the Ridley College boys' choir. Ridley had one of the top boys' choirs in North America, and as head chorister, I was the soloist. In the middle of my piece, a man placed a note on my music stand that said that my father had just died. After the concert was over, I flew back to Edmonton.

ALICE PAYNE I spent more time with my father than my brother or sister. Dad took me everywhere, and I listened to his tales over and over again. If Dad went hunting, I went along. I carried the lunch of Polish sausage or Pepper's pork pie, bread, cheese, and beer. Like the Wheeler boys, I developed the art of building duck blinds out of straw. I watched as Dad planned his next prospecting trip, and spent nights and weekends plotting Rycon drill holes on his glass model. No wonder I turned out to be a geologist. The only thing that really bothered me was the cigarette smoke.

When he was seventy, and I was twenty-one, we took what would be his last trip into the bush. Before going, he bought me a string of pearls, which my mother thought was a suitable present for a girl. Then we flew up to Yellowknife to learn the prospecting business the proper way. We hired a bush pilot, and a diamond driller, and landed near an island in the bay. We had his old canvas tent, and a well-stocked grub box. He was determined to teach me what he believed to be one of the most important things in life: staking a claim. At the time, geology lessons at university didn't include such useful information.

Dad just couldn't sit still, except in the duck swamp. His motto was, "It doesn't matter what you do, but do something, and do it well." He hoped for the best for all of his kids, and tried to emphasize honesty and truth. He often stressed the importance of keeping a few dollars aside, reminding us that he could have owned Yellowknife for fifty dollars but didn't have the cash. "Money isn't everything," he said, "But a little when you're in need can buy a lot of happiness."

# Biographies of Speakers

R.J. "Bob" Armstrong became the first mine superintendent at the Con-Rycon Mine in Yellowknife and watched the initial gold brick being poured on September, 1938. He worked at many of Cominco's mines before becoming Vice-President of the company in Vancouver. Bob and Marion Armstrong live in Penticton, British Columbia.

Ralph Cameron was a skinner, trapper, and truck driver who worked with Tom Payne on a crew for the Ryan brothers on the Smith-Fitzgerald portage. His wife Laura cooked in Joe Lanouette's Mackenzie Hotel in Fort Smith, and they spent the winters at the Halfway House.

Frank Camsell, a prospector for gold and uranium, lived at Uranium City, Saskatchewan.

Norman "Ted" Cinnamon was part of a family of six boys who worked at a variety of mining and lumber camps all over Western Canada and the North. He was at Great Bear Lake at the same time as Tom Payne, and tried to revive the sawmill after Murphy Services went bankrupt. Later, Norman lived in Yellowknife.

Frank Clune, a globe-trotting author, met Tom Payne in 1950 while stopping off in Edmonton.

Robert E. "Bob" Folinsbee, O.C., F.R.S.C., P.Geol, first met Tom Payne in Yellowknife when he worked as Jolliffe's field assistant. Later, as head of the Department of Geology at the University of Alberta, he became Alice Payne's mentor, and later her father-in-law.

John Gardner owned an antique and art shop near Buckingham Palace in London. As a young blood, he helped Tom Payne put on the Bachelor's Ball in Colchester, Essex, and years later entertained Tom and Olga when they visited England just before the Second World War.

Stanley "Stan" Graber worked on the Scottish Co-op Big Farm at Hughton, Saskatchewan in 1923, where he ran the office, did the accounting, and acted as yard foreman. His wife, aged 92 years, still lives in their home in Saskatoon, Saskatchewan.

Frank Gummerson began his lifetime of employment with Hudson Bay Mining and Smelting Co., Limited in 1926, and retired to live in Flin Flon, Manitoba.

H.W. "Harry" Hayter was a bush pilot who first encountered Tom Payne at Great Bear Lake, and was invited, years later, to attend the champagne party when Tom's well blew in. He became a Member of Canada's Aviation Hall of Fame in 1973, and died a year later at Edmonton, Alberta.

Dr. Abraham Hurtig interned at the Royal Alexandra Hospital in Edmonton, and later moved to Ottawa, Ontario, where he set up his medical practice in Obstetrics and Gynecology.

W.G. "Bill" Jewitt, one of Canada's original bush pilots, learned to fly during World War One and later graduated as a mining engineer from the University of Alberta. He began a long career with Cominco in 1927 and as an executive, he participated in managing many of Cominco's major properties. He died in 1978.

A.W. "Fred" Jolliffe, PhD., P.Geol., started out as a geologist with the Geological Survey of Canada in 1928, and after the war continued his career as a University Professor at McGill and Queen's. Jolliffe Island in Yellowknife is named after him, because the seven-tent Geological Survey base camp was struck there during 1937. He died in 1988 at the age of 81.

Ralph Keeping was the son of J.A. Pawsey's daughter, Nell. He attended Felsted School, in the family tradition, and later lived at Wombwell Hall in Barnsley, South Yorkshire, England.

Harold Kesner was a stockbroker in Edmonton and was involved with Tom Payne in the oil business. His wife, Nina, tried to teach Tom's children how to play the piano.

John Larsen immigrated to Canada from Norway and made his living as a driller. He lived most of his life in Yellowknife, in a small shack in the old School Draw. He knew Tom Payne very well, and they often enjoyed a case of Tuborg beer together. John insisted that he knew a lot about Tom that could not be printed in any book.

Ruth Maclennan is the daughter of Donald and Carrie Pawsey of Edgerton, Alberta. She took nursing training and then married Dr. Havelock Maclennan, a specialist in Obstetrics and Gynecology. She raised a family of three, and now lives in Edmonton, Alberta.

M.C. "Stan" McCormick emigrated from the United States with his family to become a farmer near Dodsland, Saskatchewan. He often hired Tom Payne to run his threshing crew in the fall.

Charles McCrea, a faithful friend of Tom Payne's, had a brilliant career as legal counsel, business executive, corporate director, and cabinet minister. He was President of Negus Mines Limited, which eventually became part of the Con-Rycon Mine in Yellowknife.

Jack Moar was a bush pilot with a degree in Mechanical Engineering from McGill University. He pioneered night airmail flights, and was named a Member of Canada's Aviation Hall of Fame in 1973 before his death four years later.

J.A. "Jim" Pawsey of Colchester, England, was the eldest of eleven boys and two girls. Three of his brothers immigrated to Canada, and settled near Edgerton, Alberta. His daughter Rosemary married Bill Payne, Tom's brother.

David Cobham Payne is Tom Payne's nephew, the son of Bill Payne and Rosemary Pawsey. He lives with his wife, Judy, at their private hotel, D'Isney Place in Lincoln, England. Their home, complete with a Roman ruin in the back yard, is adjacent to the cathedral. He is a Chartered Surveyor.

Elizabeth Pamela Payne, sister to Alice and Tommy, is an artist who lives in Edmonton. She obtained her BA in Fine Arts at the University of Alberta and a post-graduate degree at San Miguel University in Mexico.

Olive Amelia "Olga" Banks married Tom Payne. She was the daughter of Nettie Victoria Stone and William Cook Banks, who homesteaded at Forestburg, Alberta, in 1904, where they built a sod shack just in time for the birth of their first child. Olga still lives in Edmonton, Alberta, but at the age of ninety-two, has finally given up her hobbies of golf and bridge.

"Tommy" Payne is the son of Tom and Olga, their youngest child. He is a locomotive engineer who resigned from the CPR to start the first short line railway in Canada: Central Western Railway Limited. Tom and his wife Bonnie live in Edmonton, Alberta, where he is working on starting up yet another railway company.

J.J "Joe" Rankin began prospecting in northern Ontario and Quebec, and later investigated many of the same areas where Tom had been. A well-known mining man in Toronto, he worked with Mining Corp and Conwest Exploration Company, a major shareholder in Ryan Gold Mines Limited. Joe was active in the Prospectors and Developers Association and the Canadian Institute of Mining, Metallurgy and Petroleum (CIM).

M.L. "Mickey" Ryan was born in Muncie, Indiana and immigrated to Canada in 1910. He started a mail route in 1916 with his brother Pat, operating from Athabasca to Fort McMurray. The Ryan Brothers Company was successful until 1946, when it was sold. Mickey lived for a short time in Edmonton, and then moved to Sherbrooke, Quebec, where he lived until his death in 1960.

Lyman Skinner went to Great Bear Lake as a student looking for a summer job. He was not a geologist, but later worked for Fred Jolliffe at Yellowknife. He did not see Tom Payne again until many years later, when Tom turned up in Ottawa and invited him out to dinner.

Annie Smith was Tom's nurse and housekeeper for the Payne family for over fifty years. When Dr. Frank Payne died in 1947 she went to live in an almshouse in Chelmsford, Essex.

Dick Smith was a cattleman in charge of Gerald Strutt's award-winning Friesian herds in Essex. After a brief stint in Mesopotamia in World War One, he lived in Terling, Essex, most of his life.

Alan Snell was a stockbroker who lived in Edmonton, and met Tom Payne during his stint as an oilman. He was a partner in Metro Oil and Gas Limited.

Jack Stevens was a prospector in Yellowknife who was involved with Tom Payne in a group of claims north of Yellowknife known as the 'Mon Group'. He and his wife Betty were lifelong friends with Tom and Olga.

Victor K. "Vic" Stevens was a prospector who first heard of Tom Payne in the winter of 1929. Working with Dominion Explorers, he was stationed at Tavani on the west coast of Hudson Bay, and he later met Tom in Yellowknife. He retired to live at Beaverton, Ontario.

Gladys Taylor was the wife of John "Jack" Taylor, who had known Tom Payne since they were schoolmates at Felsted. She went to live with Jack at Fort Smith in a log cabin rented for ten dollars per month. She was the first female employee of the Hudson's Bay Company.

Leo Telfer was a mining engineer for Cominco, and brought his wife Dorothy and their three children to live at the Con Mine camp in 1940. Leo began exploring in the north in 1929, and first met Tom Payne in 1937 when he was a typical ragged prospector and had not yet hit the jackpot.

John Anderson Thomson, a long time resident of Yellowknife, was a surveyor and at one time, Justice of the Peace. He was always pleased to loan his canoe to the University of Alberta geology students.

Bert Wheatley came as a boy to Canada from England in 1913 to live with his brothers who were homesteaders. After serving in the army in World War One, Bert worked in Edmonton and on the west coast, then returned to Plenty, Saskatchewan, in 1925 where he met Tom.

Ken Wheeler is one of the three sons of Nell and Ben Wheeler. He obtained a B.Sc. and B.Comm. from the University of Alberta. After a career in the business world, he retired from Imperial Oil Limited and moved to Ladysmith, B.C. with his wife, Barbara. Every fall, he returns to Alberta to go duck shooting with his Pawsey relatives, reliving his experiences with Tom.

Nellie Rose Wheeler is a daughter of Donald Pawsey of Edgerton, Alberta, one of the Canadian branch of the Pawsey family. She married Dr. Benjamin Wheeler and raised three sons and a daughter, and is now married to Joe Homer. Nellie is a terrific piano player, and at the age of 89, can still pound out the St. Louis Blues for her visitors in Edmonton.

Gordon Willsie was a prospector in northern Manitoba for much of his life, and retired to live in Thetford, Ontario. He donated his extensive photo collection to the National Archives of Canada in Ottawa.

Steve Yanik lived in the north most of his life, and was a prospector who followed Tom's pattern of success. Steve first met Tom Payne at Great Bear Lake, where he worked for Eldorado and mucked out the first round of ore taken from the mine. He made his fortune when he staked his claims at Pine Point next to Cominco. He and his wife Irene live in Edmonton, Alberta.

# General Notes

I have chosen to retain the terms Eskimo and Indian contained within the older documents in order to retain their historical context. Where appropriate, the terms Inuit and Dogrib (Dene) have been substituted to reflect modern times. Historical place names should be noted as follows: Fort Franklin (Deline), Fort Norman (Tulita), Eskimo Point (Arviak) and Coppermine (Kugluktuk.).

Cominco was once officially known as The Consolidated Mining and Smelting Company of Canada, Limited. This long name was frequently abbreviated to 'Consolidated Smelters' or 'Smelters' or simply 'Con' or 'C.M. & S.' and finally 'Cominco'.

# Selected Bibliography

Alcock, F. J. *A Century in the History of the Geological Survey of Canada*, Ottawa: Geological Survey of Canada Special Contribution 47-1, 1947, pp. 36-40.

Beck, Carl. "Sage Comment: Making a Prospector – Gordon Willsie." *The Northern Miner* (31 May 1973): 4.

Boyle, R.W. "An Occurrence of Native Gold in An Ice Lens – Giant Yellowknife Gold Mines, Yellowknife, Northwest Territories." *Economic Geology* 46 (1951).

Campbell, Neil. "The West Bay Fault, Yellowknife." *Western Miner* 20 (1947): 46-55, Part 1.

Canadian Institute of Mining and Metallurgy. "Structural Geology of Canadian Ore Deposits." *CIM Symposium Jubilee Volume, Montreal, QC.* (1948): West Bay Fault, 244-69; Eldorado Mine, 259-268; and Sherritt Gordon Mine, 292-301.

*Canadian Mining Journal.* "Hudson Bay Mining & Smelting Co. which details a Great Transportation Job and Latest News From the Mine." 50 (1929).

Clune, Frank. *Hands Across the Pacific, A Voyage of Discovery from Australia to the Hawaiian Islands and Canada, April to June, 1950.* Sydney, London: Angus and Robertson, 1951.

Craze, Michael. *A History of Felsted School,* 1564-1947. Ipswich, UK: W.S. Cowell Ltd., 1955.

*Essex County Chronicle* "Funeral of Essex Man Known for Generosity." (3 February 1950). [Portrait of H.M. Mont Everard.]

Duchaussois, Father Pierre. *Mid Snow and Ice, the Apostles of the North-West.* Buffalo, NY: Missionary Oblates of Mary Immaculate, 1937.

*Edmonton Journal.* "Six Men Finish 100-Mile Trek Over Barrens." (10 November 1933.) [The Speed II disaster in 1933.]

*Edmonton Journal.* " Prominent Mining Men Enthusiastic Over Find High Grade Pitchblende. Claim to Have "Richest Radium Strike" in Great Bear Lake Area." (19 March 1934). And "100,000 Pounds Radium Ore At $400 to $1,750 Per Ton To Be Flown Out of North. Edward Hargreaves of Toronto, Announces Contract Signed." (21 July 1934). [Interview with E.H. Hargreaves Sr. and E.H. Hargreaves Jr.]

*Essex Weekly News* (25 November 1898). "A Witham Doctor Prosecuted." [Payne family lawsuits. See also *Essex County Chronicle* ."Alleged Assault by a Witham Doctor." (25 November 1898). *East Anglian Daily Times.* "Witham Doctor Summoned." (23 November 1898).]

*Financial Times.* "Southern Alberta Land Co." (23 July 1914).

Finnie, Richard. *Canada Moves North.* Toronto: MacMillan Co. of Canada Ltd., 1942.

Fleming, H.A. *Canada's Arctic Outlet: A History of the Hudson Bay Railway.*" Berkeley: University of California Press, 1957.

Gard, Robert E. *Johnny Chinook.* Published by Longmans, Green and Company, 1945.

Gavin, Sir William. *Ninety Years of Family Farming: the Story of Lord Rayleigh's and Strutt and Parker Farms.* London, UK: Hutchinson of London, 1967.

Graber, Stan. "Muddy Roads and Cold Didn't Stop Our First Cars," *Grainews* (June 1993): 50. And "Big Farms in Early Saskatchewan: 1923 Changed from a Drought to Produce a Bumper Crop." (12 April 1993): 49.

Hayter, Ron. "North Pays Tribute to Yellowknife Man." *Edmonton Journal* (11 December 1961). [See also Ed Cosgrove. "Vic Ingraham Dies. 'Indestructible' Heart Stops." *The Daily Colonist* (15 November 1961): 13.]

Hedman, Valerie and L. Yauck and J. Henderson. *Flin Flon.* Altona , MB: Flin Flon Historical Society, 1974.

Hoffman, Arnold. *Free Gold: The Story of Canadian Mining.* NY: Associated Book Service, 1947.

Jackson, James A. *A Centennial History of Manitoba,* Toronto: McClelland and Stewart Ltd., 1970.

Jackson, Susan. ed. *Yellowknife, N.W.T.: An Illustrated History.* Volume 1 Yellowknife History Series. Sechelt, BC: Nor'West Pub., 1990.

Jolliffe, A.W. *Geological Survey of Canada Field Notes.* Ottawa: Geological Survey of Canada, book 268 (1) 1934.

Jolliffe, A.W. "A Peek Into Yellowknife's Past – By One Who Was There." *The Northern Miner* (28 November 1974): 55-56.

Jolliffe, A.W. "Yellowknife River Area, Northwest Territories, 1936." And "Yellowknife Bay – Prosperous Lake Area, Northwest Territories, 1938." Preliminary Series Papers 36-5, and 38-21. Ottawa: Geological Survey of Canada.

Kidd, D.F. and Haycock, M.H. Mineralogy of the Ores of Great Bear Lake. *Geol. Soc. of America. Bulletin* 46 (1935):879-960.

Lord, C.S. "Mineral Industry of the Northwest Territories." Ottawa: Geological Survey of Canada Memoir 230, 1941. [For reports on mineral properties.]

Low, A.P. *The Cruise of the Neptune*, Report on the Dominion Government Expedition to Hudson Bay and the Arctic Islands on Board the DGS Neptune, 1903-1904. Ottawa: Government of Canada, 1906.

MacEwan, Grant. *West to the Sea*, Scarborough, ON: McGraw-Hill, 1968, pp. 53-58. [The Palliser Expedition.]

MacEwan, Grant. *Eye-Opener Bob.* Edmonton: Institute of Applied Arts, 1957.

MacInnes, Tom. *Klengenberg of the Arctic: An Autobiography*. Toronto: Jonathan Cape Ltd., 1932.

Moar, Kitty. *A Collection of Bush Flying Stories.* Victoria: Limited Edition, 1991.

Morton, W. L., *Manitoba: A History*. Toronto: University of Toronto Press, second edition, 1967.

Nagle, Ted and Jordan Zinovich. *The Prospector north of sixty*, Edmonton, AB: Lone Pine Publishing, 1989.

Nikic, Z. and H. Baadsgaard, R.E. Folinsbee, J. Krupicka, and A. Payne Leech. "Boulders from the Basement, the Trace of an Ancient Crust?" *Geological Society of America*, Special paper 182, 1980.

*Northern Miner*. "Prospectors Fight Polar Conditions in Record Flight." (25 April 1929.) [MacAlpine Expedition.]

*Northern Miner.* "Hudson Bay Mining." (26 July 1973):1.

*Northern Miner.* "Thayer Lindsley, Founded Falconbridge, Many Other Mines." (3 June 1976): 6.

Oswald, Mary E. ed. *They Led the Way: Members of Canada's Aviation Hall of Fame*. Wetaskiwin, AB: Canada's Aviation Hall of Fame, 1999.

*Ottawa Citizen.* "Thrilling Tale of Battle For Life In Gale on Bear Lake. (13 November 1933.) [The Speed II disaster in 1933.]

Pennie, A.M. "A Cycle at Suffield." *Alberta Historical Review* vol. 11, 1 (1963): 7-11. [Irrigation, and the Southern Alberta Land Company].

Pryde, Duncan. *Nunaga: My Land, My Country.* Edmonton: M.G. Hurtig Ltd., 1971.

RCMP Annual Report, 1925, p. 45. [For an account of the death of Sergeant S.G. Clay's wife by sled dogs.]

Schmidt, John. *Growing Up in the Oil Patch.* Toronto: Natural Heritage/ Natural Heritage Inc., 1989.

Shepard, George. *West of Yesterday.* Toronto: McClelland and Stewart, 1965.

Stefansson, Vilhjalmur. *My Life with the Eskimo.* New York: First Collier Books, 1962.

Stockwell, C.H. and Kidd, D.F. "Metalliferous mineral possibilities of the mainland part of the Northwest Territories." Ottawa: Geological Survey of Canada Summary Report 1931. Part C, 1932.

Tourism Committee Fort Smith. *On the Banks of the Slave: A History of the Community of Fort Smith, Northwest Territories.* Education Programs and Evaluation Division. Department of Education, NT. 1979.

Tranter, G.J. *Link to the North.*, London: Hodder and Stoughton Ltd., 1946. [The story of the Ryan brothers.]

Treilhard, D.M. (Mac). Fire and Ice: Ordeal at Great Bear. *Up Here* (August-September 1986): 36-39. [The Speed II disaster in 1933.]

# Index

## A



## B

**C**

## D

**E**



**F**

**G**

## H



## I

## M

## N



## O



## P

## Q

## R



## S

**Y**